Fundamentals of Free Space and Deep Space Optical Communications

Shun-Ping Chen

Fundamentals of Free Space and Deep Space Optical Communications

Models, Components, and Link Performance

Springer

Shun-Ping Chen
Fachbereich EIT
Hochschule Darmstadt
University of Applied Sciences
Darmstadt, Hessen, Germany

ISBN 978-3-032-35825-7 ISBN 978-3-032-35826-4 (eBook)
https://doi.org/10.1007/978-3-032-35826-4

This Springer imprint is published by the registered company Springer Nature Switzerland AG
The registered company address is: Gewerbestrasse 11, 6330 Cham, Switzerland

To Michael, Anna, and Katja

Preface

This book is designed for both the scientific audience, research and development engineers and also university students. It explains the fundamentals of free space and deep space optical communications by considering different applications, system components, link distances, and the real-world conditions like atmosphere, cloud, and fogs.

Up to now the classical microwave radiowaves are mainly utilized for the free space communication. Radiowaves have a long history, starting with the demonstration of electromagnetic wave propagation by Heinrich Hertz (1886) and the invention of radio frequency communications by Guglielmo Marconi (1895).

A large range of free space or wireless communication applications like radio and TV broadcasting, mobile communications, satellite communications, and deep space communications for space science projects based on radio frequencies have been developed rapidly, and still dominate most wireless communications.

In the last decades, also due to the rapid, successful development of fiber optical communication technologies since 1960s, with many types of high-power transmitters and photon sensitive receivers in the last decades, free space optical communication (FSOC) is no more only used for short-distance terrestrial communications and inter-satellite links, but also investigated and designed for deep space optical communications (DSOC) demonstration projects. DSOC shows especially significant advantages and potentials in terms of compact transmitters and receivers, lower weight on the spacecraft, small beam spot size, therefore larger wave propagation distance and higher data rates. All these advantages can improve space communication technologies.

This is also the motivation for international space agencies like NASA and ESA to successfully conduct the Lunar Laser Communication Demonstration (LLCD) from October 2013 to April 2014 and several other DSOC projects like the demonstration project Psyche Mission of NASA in cooperation with ESA which started in October 13, 2023. The first Psyche DSOC measurement results were successfully received at the ground station end of 2023 and beginning of 2024, verifying and confirming the high potentials of DSOC in comparison with classical radio frequency communications. In 2026 the Artemis II mission to the Moon successfully used an optical link, O2O, to transmit high resolution videos from space. Pulse-Position Modulation (PPM) schemes (16-PPM and 32-PPM) were employed to achieve high data rate transmissions over these large distances.

Even though the FSOC and DSOC will not replace the radio wave propagation completely in the future, they will be definitely promising candidates and successful for certain applications.

Beginning with the FSOC/DSOC system model and basic components of the general free space optical communication channel, the most important effects affecting free space optical wave propagation are introduced. The required optical power is basically depending on the distance between the transmitter and receiver. The received optical power continuously decreases with increasing distance between transmitter and receiver due to the increasing path loss or beam divergence. If the optical wave propagates through the atmosphere, i.e. communication links between Earth ground stations and satellites or spacecraft, then the absorption and scintillation in the atmosphere will affect the wave propagation additionally, leading to additional loss and dynamic variation of the signal strength.

We distinguish between terrestrial, near-Earth, lunar and deep space optical communications. By considering the absorption or attenuation and the scintillation in the atmospheric layers, the link performance of free space optical communications and deep space optical communication has been investigated and will be discussed more in details in this book.

This book focuses on the link performance analysis of FSOC, DSOC and telecommunication technologies. It also covers projects the author was involved in during the last years, both at Darmstadt University of Applied Sciences and as visiting research scientist with ESA/ESOC. The design aspects of spacecraft, tracking, pointing and acquisition, navigation, control mechanisms, encryption techniques and quantum key exchange in free space optical communications will not be content of this book. There exist some standard books and papers published by NASA JPL specialists and ESA scientists with rich, hands-on experiences also listed in the references in the respective chapters in this book.

The author is very happy and lucky that in the last few years up to actually several weeks and days ago, missions like LLCD, Psyche, and Artemis II conducted by NASA in cooperation with ESA have achieved successful demonstration project results which can be compared with the simulation models developed in the past and confirm the suitability of these models. These results are presented in this book.

The author hopes that this book, which is more an overview of basic concepts and fundamentals of FSOC and DSOC, can motivate university students and young research engineers to familiarize themselves with these new technologies, and possibly extend the knowledges by reading other in-depth research papers and books cited in the references of corresponding chapters in this book.

Darmstadt, Germany
June 2026

Shun-Ping Chen

Acknowledgements The author would like to thank all colleagues of the Institute of Communication Technologies and the Department of Electrical Engineering and Information Technologies, Darmstadt University of Applied Sciences (h-da), for enabling and supporting the research activities, i.e. research sabbaticals at ESA/ESOC with the involvement in Deep Space Optical Communications and the Radio Frequency Amplitude and Phase Scintillation projects, and for many helpful discussions.

The author firstly contacted ESA/ESOC in Darmstadt and visited with a group of students of Communications Major the Ground Station System Division Director Dr. Klaus-Jürgen Schulz and Section Head Dr. Igor Zayer in 2018 who introduced us to the exciting field of Deep Space Optical Communications (DSOC) and Free Space Optical Communications (FSOC), and also showed us their laboratories. This was the starting point to get more and more involved in the research activities in DSOC and FSOC, especially in the analyses and theoretical simulations of FSOC and DSOC link performance. Teaching activities of the author are lectures in Mobile Communications, Wireless Systems and Technologies, Fields, Waves, and Antennas, Optical Communications, Digital Technologies or Logic Design, Fundamentals of Electrical Engineering etc.

In parallel to some research activities and publications, the author also proposed and guided some student research projects in terms of literature survey on FSOC and DSOC, and basic simulations, in order to motivate them to learn more about these new, fascinating technologies and how the research activities could generally look like in their future career.

These first research activities, also together with his students, motivated the author extremely, so that he applied for two research sabbaticals from October 2020 to October 2021 with ESA/ESOC.

The author would like to thank ESA/ESOC and Darmstadt University of Applied Sciences h-da for the extremely valuable opportunity to be a Visiting Scientist with ESA/ESOC Ground Station Division in 2020–2021 involved in two very fascinating research projects of (1) Deep Space Optical Communications DSOC and (2) Investigation of X- and Ka-Band amplitude and phase scintillation during the superior solar conjunction. He especially appreciates many inspiring discussions with Dr. Clemens Heese, Dr. Robert J. Daddato, Dr. Zoran Sodnik, Dr. Marco Lanucara, Dr. Jose Villalvilla, and other colleagues at ESA.

He would like to thank all the students, Tobias Siegel, Emmanuel Domfeh Aboagye, Wassim Messaoudi, Sana Guezguez, Katharina Sauer, Bangalore Vijayananda Sandeep Sharma, Patel Sukesh, Felix Schüßler, Tobias Stieglitz, Patrick Thielke, Rithika Grace Robert Charles, Sowmiya Sukumaran, Vaiby Jose, Vinay Sadashivpet, Neha Manjunath, Venkatesh Dharamkari, Mounika Bharadwaj, Adrian Alexander Maurer, Omar Sobhy Ibrahim Shoeib, Saurabh Srivastav, Muhammad Adil, Saraha Shakir, Rutvik Nasit, Sai Pramitha Meesala, who joined the research projects in FSOC and DSOC, proposed and guided by the author in the last few years, for their dedication and contributions with very interesting results.

The author also would like to thank Friederike Lierheimer, Sonam Yadav, and Shirley Regina D. of Springer Nature for their support during the book production process.

Finally the author appreciates the continuously good cooperation and many fruitful discussions with his friend, colleague, and consultant Prof. Dr. Heinz Schmiedel about Microwave Engineering and Optical Communications, for good cooperation in research, careful proofreading of all the drafts of my books and papers prior to submission, and for many valuable proposals.

Competing Interests The author has no competing interests to declare that are relevant to the content of this manuscript.

Contents

1 Introduction, Motivations

Up to now, mainly classical microwave radiowave bands are utilized (see Appendix) for free space communications. Historically it goes back to the demonstration of electromagnetic wave propagation and the invention of antennas by Heinrich Hertz and Guglielmo Marconi.

A large number of free space or wireless communication applications like radio and TV broadcasting, mobile communications, satellite communications, and deep space communications for space science projects have been developed rapidly.

In the last decades, also due to the rapid development of fiber optical communication technologies since the 1960s with many successful transmitters and receivers, free space optical communication (FSOC) Free Space Optical Communications is not only used for short-distance terrestrial communications, but also gradually investigated and implemented for deep space communications. FSOC shows significant advantages in terms of compact transmitters and receivers, lower weight on the spacecraft and small spot size in the research pilot or demonstration projects. These are the motivations for international space agencies like NASA and ESA to successfully conduct several deep space optical communication (DSOC) projects. Even though the FSOC and DSOC will not replace the radio wave propagation completely, they will be definitely promising candidates and successful for certain applications.

The most important milestones of history of telecommunications from ancient methods to exchange information by using acoustic signals, fire signals, and optical telegraphs to modern electrical systems are summarized in [1, 2].

S.-P. Chen, *Fundamentals of Free Space and Deep Space Optical Communications*,
https://doi.org/10.1007/978-3-032-35826-4_1

1.1 Prehistoric Communications with Fire Signal and Optical Telegraphs

4th century—In ancient Greece, the hydraulic semaphore was built and used for communications. It played a vital role during the first Punic War.

5th century—Persia–Syria established the first pigeon-short-messaging system after the discovery of the uncanny ability of the pigeons.

6th century—The first postal system established by Cyrus was adopted later-on by the other ancients (Egypt, Rome, and China) with incorporating slightly new mechanisms.

1666—Flag semaphore originated as a handheld version of the optical telegraph system of Home Riggs Popham used on land, and its later improvement by Charles Pasley. Specific codes displayed by positioning the hand-held flags to communicate between ships was established. Every movement of the flags indicated an alphabet or numerical value.

1672—Robert Hooke first demonstrated an acoustic telephone and proved that acoustic signals can travel over wire from a microphone to a speaker.

1790—The Chappe brothers established the first optical telegraph system in Europe referring to the naval flag semaphore. It was a pendulum-like system built on a tower. The telegraph system had mechanical arms that swung around to convey messages between tall towers.

1838—Samuel B. Morse had been working with his friends Alfred Vail and Leonard Gale on the idea of a recording telegraph. They established a push-button-based message transmission system between two telegraphs connected together with a pair of current carrying wires. This technique is now called Morse-code and it laid the base for several wired communication systems.

1858—At this time, telegraph stations for continuous communications were installed in Britain and the United States. However laying a telephone cable to establish telephonic communication between England and the United States was the wish of Cyrus Field in New York.

1861—Johann Phillipp Reis from Friedrichsdorf near Frankfurt in Germany invented the first working telephone which transmits music and speech over an electrical wire.

1866—Flag semaphore originated as a handheld version of the optical telegraph system of Home Riggs Popham used on land, and its later improvement by Charles Pasley. Specific codes displayed by positioning the hand-held flags to communicate between ships was established. Every moment of the flags indicated an alphabet or numeric value.

1867—In this year, Adml. Phillip introduced a lamp-based code (dots/dashes) signalling techniques at sea for establishing communication between ships.

1.2 Beginning of Modern Communications with Major Development Milestones

1876—This was a great year for A. Graham Bell. With an assignment of a teacher (in a deaf school) in the USA, he was researching to build an electronic system to transmit speech signals which eventually was the birth of the telephone.

1877—Thomas Alva Edison built an acoustic phonograph to record/transmit sound signal.

1880—A. Graham Bell built a laboratory and demonstrated his invention 'photophone' through which he transmitted a sound signal using a beam of light. That is the first wireless call, so A. Graham Bell owns a place in optical communications history. A historical plaque, placed on the side of the Franklin School in Washington, D.C marks one of the points on which the photophone was demonstrated. This photophone demonstration was the first take for today's free space optical (FSO) communication technology.

1893—Nikolai Tesla transmitted radio waves wirelessly through a transmitter (TX).

1896—Marconi sent his radio waves over a distance of 2 km.

1915—A. Graham Bell was credited with an entry in the history book second time due to his long land-line phone at a coast.

1927—Phillip T. Farnsworth demonstrated the first working television set. The wireless telephone service was made available between the UK and the USA.

1930—It was decided by 'American Telephone and Telegraph' to have a demonstration videophone.

1934—First wireless telephone calls were made available between USA and Japan.

1936—First videophone connectivity was made in Germany for the public use and exhibited it in a trade-fair.

1946—First automobile voice telephone call was demonstrated.

1956 – Newfoundland and Scotland got connected via a transatlantic voice telephone cable with 36 circuits.

1950s—Charles Townes and Arthur Schawlow advanced concepts, leading to the invention of the "MASER" (microwave amplification by stimulated emission of radiation), the precursor of the Laser.

1962—The wireless satellite communication act was officially passed and it enabled the establishment of wireless services through space.

1.3 Start of Fiber Optical Communications and Free Space Laser Optical Communications

1960—On May 16, 1960, the American physicist and engineer Theodore Maiman at the Hughes Research Laboratories created the world's first functioning laser (often written as LASER). Charles H. Townes, Nikolay Basov, and Alexander Prokhorov later received the Physics Nobel Prize for the invention of the laser.

1962—IBM's GaAs semiconductor laser diode really ignited the laser revolution.

1964—Charles Kao and George Hockham proved the possibility of voice communication using a light signal sent through a fiber optical cable. They published their experimental results and findings in a paper which actually revived the photophone of A. Graham Bell. For this invention Charles Kao receved the Physics Nobel Prize in 2019.

1960s—Invention of semiconductor laser diodes by Zhores I. Alferov and Herbert Kroemer was later awarded the Physics Nobel Prize "for developing semiconductor heterostructures used in high-speed and opto-electronics". Their work in the 1960s allowed the first continuous-wave operation of semiconductor lasers at room temperature, which made fiber-optic communications, CD players, and barcode scanners possible.

1965—First picture-phone service began in trials.

1969—Digital data travelled for the connectivity of ARPANET's nodes and the Internet's predecessor.

1973—First cellular mobile voice call was made, for the talk time of 30 min, by Inventor Martin Cooper at Bell Laboratories and the mobile phone battery took a year to get fully charged.

1979—As per the decision and requirement of INMARSAT, maritime vessels were connected by reliable, safe and secured communications.

1981—Japan got the commercial automated cellular network deployed.

1982—Highly secure and limited internet cluster access was given to the military, corporate and research facilities of few universities.

1983—ARPANET standardized internet connectivity and service providing protocols through network control protocols and transmission control protocol/internet protocol (IP).

1998—Iridium put 64 satellites into orbit, mainly for phone communications.

2003—IP-assisted phone calls were made using the existing computer internet network.

Beginning with invention of lasers (1960), GaAs laser diodes (1964), and the optical fiber [3] in 1964s, the modern fiber optical communication system components like signal processing unit, modulator, transmitter, receiver, and demodulator can be either directly used or modified for free space optical communications (FSOC) or deep space optical communication (DSOC).

1.4 The Important FSOC and DSOC Projects

The most important FSOC and DSOC projects are [4–10]:

1970s—ESA started laser communication technology development in the mid of 1970s [6].

2001—ESA demonstrated the world-first optical inter-satellite communication link (SILEX) [6] between the ARTEMIS data-relay satellite in geostationary orbit (GEO) and the SPOT-4 Earth observation satellite in low earth orbit (LEO).

2007—German Space Agency DLR developed the Laser Communication Terminals (LCT) and launched LCT on 2 LEO Spacecrafts (TerraSAR-X and NFIRE), later between LEO and GEO.

2013—NASA's Lunar Atmosphere and Dust Environmental Explorer (LADEE) satellite's main mission goal is to study the pristine state of the lunar atmosphere and dust environment, with Lunar Laser Space Terminal (LLST) for the demonstration of high speed (up to 622 Mbps downlink and up to 19 Mbps uplink) optical communications with a ground station.

2013—MIT coordinates the Lunar Laser Communication Demonstration (LLCD) Project, and invited the Jet Propulsion Laboratory (JPL) and European Space Agency (ESA) to participate with their own ground facilities: Optical Communication Telescope Laboratory (OCTL) on Table mountain, California and Optical Ground Station (OGS) in Tenerife.

2023—Psyche Mission was launched on October 13, 2023 in cooperation with ESA's with a Deep Space Optical Communications Demonstration. The first measurements results have already been achieved.

2026—From April 1, 2026 - April 11, 2026, during NASA's Artemis II mission, partially in cooperation with ESA, a laser-based system called O2O (Orion Artemis II Optical Communications System) onboard ORION spacecraft was used to transmit data with a rate up to 260 Mbit/s.

Keeping the historical development and the successful missions of FSOC and DSOC in mind, this book begins with the model and basic components of a general free space optical communication channel. Then the most important effects affecting free space optical wave propagation are introduced.

The optical power is basically dependent on the distance between transmitter and receiver. The optical power will be continuously reduced with increasing distance due to the path loss or beam divergence. If the optical wave propagates through the atmosphere, i.e. the communication links between Earth ground stations, satellites or spacecrafts, absorption and scintillation in the atmosphere will affect the wave propagation, leading to additional loss and variation of the signal strength which is called scintillation.

We distinguish between terrestrial, near-earth, lunar and deep space optical communications. By considering absorption or attenuation and scintillation in atmospheric layers, the link performance of free space and deep space optical communication have been investigated and are discussed in this book.

All the electromagnetic wave propagation problems can be described generally by Maxwell's equations, valid from L, S, C, X bands (1–12 GHz), Ku, K, Ka bands (12–18 GHz), mm-Waves (100–300 GHz), THz band (300–3 THz) to optical frequencies like infrared (3–480 THz) and visible light (480–680 THz).

By using vector algebra, Maxwell's equations can be derived in different coordinate systems depending on the special problems to be investigated, i.e. wave guidance, radio wave radiation, or generally wave propagation.

For example, for rectangular metallic waveguides, the Cartesian coordinate system (x, y, z, t) will be meaningful, whereas for problems of cylindrical metallic or quartz glass optical waveguides, the cylindrical coordinate system (r, ϕ, z, t) will be used. For the antenna problems the polar coordinate system (r, θ, ϕ, t) will be helpful.

The vector algebra and the derivation of Maxwell's equations into the Helmholtz equations as a good approximation are described more in detail for example in [11–15]. Also some wave propagation problems both for laser optical and radio frequency free space communications by considering atmospheric scintillations are discussed in [16, 17].

One of the most popular books about the deep space optical communications (DSOC), written by Hemmati [18] with all the details should be recommended to all interested scientists, engineers, and students who would like to get more involved in in-depth research activities.

References

1. A. B. Raj , A. K. Majumder: Historical perspective of free space optical communications: from the early dates to today's developments. IET (The Institution of Engineering and Technology) Commun., 2019, Vol. 13 Iss. 16, pp. 2405–2419.
2. History of Telecommunication. https://en.wikipedia.org/wiki
3. K. C. Kao, G. A. Hockham: Dielectric-fibre surface waveguides for optical frequencies. Proceedings of the IEE, Vol. 113, No.7, July 1966.
4. Zoran Sodnik, Clemens Heese, Pantelis-Daniel Arapoglou, Klaus-Jürgen Schulz, Igor Zayer, Robert Daddato, Stefan Kraft, "Deep-Space Optical Communication System (DOCS) for ESA's Space Weather Mission to Lagrange Orbit L5". IEEE International Conference on Space Optical Systems and Applications (ICSOS), November 14–16, 2017.
5. D. Cornwell, "The Lunar Laser Communication Demonstration (LLCD)". LLCD Mission Manager, NASA GSFC, Presentation to the JHU Aerospace Affinity Group. June 11, 2014.
6. Z. Sodnik, H. Smit, M. Sans, D. Giggenbach, P. Becker, R. Mata-Calvo, C. Fuchs, I. Zayer, M. Lanucara, K.-J. Schulz, J. Widmer, F. Arnold, M. Mosberger, A. Alonso, I. Montilla, "Results from a Lunar Laser Communication Experiment between NASA's LADEE Satellite and ESA's Optical Ground Station", Proc. International Conference on Space Optical Systems and Applications (ICSOS) 2014, S2-1, Kobe, Japan, May 7–9 (2014).
7. Z. Sodnik and R. Czichy, "Design Data Summary of the ESA Optical Ground Station for In-Orbit Check-Out of Laser Communication Payloads and for the Observation and Registration of Space Debris". ESA Specifications for the Artemis OGS OPALE project. Ref.: TEC-MMO/267/ZS. May 2, 2005.
8. NASA, "NASA's Optical Comms Demo Transmits Data Over 140 Million Miles". (2024), URL: https://www.nasa.gov/missions/psyche-mission/nasas optical-comms-demo-transmits-data-over-140-million-miles/ (Retrieved on May 15, 2024).
9. C. A. Polanskey et al., Psyche Mission Description and Design Rationale. Space Science Reviews (2025) 221:95 https://doi.org/10.1007/s11214-025-01218-x
10. NASA: http://www.nasa.org/technology/space-comms/o2o/
11. R. E. Collin, F. J. Zucker: Antenna Theory, Part 1. McGraw-Hill Book Company (1969).
12. H.-G. Unger: Planar Optical Waveguides and Fibres. Oxford Clarendon Press (1977).
13. C. A. Balanis: Antenna Theory. John Wiley & Sons, Inc. Fourth Edition (2016).

14. K. W. Kark: Antennas and Radiation Fields (in German: Antennen und Strahlungsfelder). Springer Vieweg (2018).
15. S.-P. Chen: Fundamentals of Information and Communication Technologies. Cambridge Scholars Publishing (2020).
16. S.-P. Chen: Investigations of free space and deep space optical communication scenarios. CEAS Space Journal, Springer Nature (2021).
17. S.-P. Chen, J. Villalvilla: Comparison of Modified Woo's Solar Phase Scintillation Model with ESA's BepiColombo Superior Solar Conjunction Measurement Data for X-Band and Ka-Band. CEAS Space Journal, Springer Nature (2022).
18. H. Hemmati (Editor): Deep Space Optical Communications. Wiley, 2005.

2 Free Space Optical Communication System Model

2.1 Introduction

Free space optical communication FSOC enables the data transmission by modulating the payload information onto the electromagnetic waves in the infrared light bands in free space. The systems consist of an optical transceiver (transmitter and receiver) at both ends to provide bidirectional capability. The attempt is very similar to optical fiber communication systems, just without the need of a dedicated quartz glass optical fiber as transmission media. The technology itself is not new, but with the invention of laser diodes and highly sensitive photodiodes as well as sophisticated modulation schemes, it now meets the specifications for the rising demand of high bandwidth connections.

There are multiple applications where the advantages of free space optical communications (FSOC) can be utilized. Unlike the limited spectrum availability with radio frequency communication, optical communication has a license-free spectrum. Since there is no fiber cable required, a connection can be established very flexibly and quickly with free visual contact or Line of Sight (LoS) between two points. This does not only enable quick connectivity after, e.g. natural disasters, but also a stable connection with moving objects like airplanes or even satellites. But the precondition for visual contact has also a downside.

The transceivers always must be adjusted very precisely to prevent pointing errors. And even if this is the case, unpredicted obstacles or bad atmospheric conditions in the line of sight can still influence the signal quality negatively. The primary factors that deteriorate the link are absorption, scattering, and turbulence. The given limitations narrow the use cases for free space optical communications in common commercial installations where a reliable connection is mostly mandatory. Due to the usage of point-to-point connections, scaling a network is more complex and requires additional maintenance for each connection. Because of the narrow laser beam, free space optical communications is much more secure than radio-frequency transmissions, but eavesdropping is still possible. Additional encryption techniques in the network layer or transport layer like IPsec or TLS/SSL are therefore

S.-P. Chen, *Fundamentals of Free Space and Deep Space Optical Communications*,
https://doi.org/10.1007/978-3-032-35826-4_2

recommended. Both encryption techniques and quantum key exchange will not be content of this book.

Generally Maxwell's equations are the governing equations to analyze all electromagnetic wave propagation problems, from RF radio waves used in cellular mobile communications 1–2 GHz, mm waves (frequency range of 30–100 GHz) up to optical wave propagation in the frequency range of 200 THz (see also [1–6]). Depending on the applications, different carrier frequencies are used from 1 GHz to mm-waves or even to the optical spectrum. These frequencies have different abbreviations, standardized by different organizations like IEEE, ITU, EU and NATO, and are listed in the Appendix A.

$$\nabla \times \mathbf{E} = -\frac{\partial \mathbf{B}}{\partial t}, \tag{2.1}$$

$$\nabla \times \mathbf{H} = \frac{\partial \mathbf{D}}{\partial t} + \mathbf{J}, \tag{2.2}$$

$$\nabla \cdot \mathbf{D} = \rho, \tag{2.3}$$

$$\nabla \cdot \mathbf{B} = 0. \tag{2.4}$$

By applying these fundamental Maxwell's equations and vector algebra to different coordinate systems, the different wave propagation or wave guidance problems can be solved [1–6], i.e. circular and rectangular waveguides, optical fibers, antenna systems, and free space wave propagation, also by considering the atmospheric scintillation effects in RF and optical frequency bands [7, 8] (Table 2.1).

In case of free space optical communications or deep space optical communications, the system design parameters can be strongly varying depending on the totally different scenarios. One of the very popular books [9] covers all system design aspects of the deep space optical communications.

By applying the above fundamental theory, different free space optical communication scenarios are investigated [8, 10–13]. The simulation for almost all investigations was mainly done by using Matlab [15] of MathWorks and OptiSystem [16] of OptiWave for short distance

Table 2.1 Electromagnetic field parameters and units

Field vector	Symbol	Unit
Electric field	**E**	V/m
Dielectric displacement/density	**D**	As/m^2
Magnetic field	**H**	A/m
Magnetic flux density/induction	**B**	$T = Vs/m^2$
Current density	**J**	A/m^2

Table 2.2 Examples for typical simulation parameters for free space optical communications (FSOC) and deep space optical communications (DSOC)

Parameter	Symbol	Value	Unit
Planck's constant	h	6.624×10^{-34}	Joules/Hertz
Electron charge	Q	1.6×10^{-19}	Coulombs
Boltzmann constant	k_B	1.38×10^{-23}	Joules/Kelvin
Load resistance	R_L	$5.75 \times 10^{12} \times Ts$	Ohm
Elevation Field of View	θ_{FOV}	20	Degrees
Signal slot width	Ts	Variable	Seconds
Laser TX power	P_{avg}	5.0	Watt
TX aperture	D_{tx}	0.22	Meters
RX aperture	D_{rx}	5.0	Meters
Link margin	L_{margin}	3.0	dB
Optical efficiency	η	0.35 for Detector	
Pointing loss	L_{Point}	2	dB
Transmission loss	L_{Trans}	0.55	
Wavelength	λ	1.55	μm
PPM order	M	Variable	
Bits per symbol	k	Variable	bits
Sky radiance	Hb	0.0085	W/(cm$^2 sr \mu$m)
Atmospheric transmittance	T_{zenith}	0.95	
Relative air mass	M_r	2.9	
Code rate	CR	1/3	
Fried parameter	r_0	0.05	Meters
Scintillation loss	L_{scint}	0.1	dB
Bit rate	BR	100	Mbit/s

FSOC by using QPSK and 16-QAM. Just as an example one use case for free space optical communications shows the typical system parameters involved in the following Table 2.2 [10].

Due to the increasing demand for higher bandwidth in modern communication systems, conventional networks are continuously expanded with new technologies to improve coverage. Free space optical communications (FSOC) shows some significant advantages concerning system setup time in comparison with the classical fiber optical systems on one hand, substantial spectral bandwidth and performances in comparison with wireless systems by using radio frequency bands under certain conditions on the other hand. This makes this technology not only a reasonable extension for metropolitan area networks but also provides the capability to set up a network quickly after an outage in case of natural disaster. But transmitting data by using FSOC involves some limiting factors that have to be considered prior to each installation.

Since the atmospheric channel is not static, the influence of changing weather conditions or industrial smog have a significant impact on the available bitrate. A simulation platform is developed and presented in this book for investigation of FSOC considering these circumstances. Regarding the atmospheric channel, turbulence, distance-dependent beam divergence, and applied modulation schemes, a general overview of the capabilities is presented and discussed. The insight of this book should help to make a decision under which preconditions either the FSOC provides a meaningful application possibility, or the limiting factors become too crucial and other technologies must be considered.

2.2 Shannon Channel Capacity Limit

A digital signal can be transmitted in a channel defined by its bandwidth (in Hertz) and by the noise and unwanted interference from other users. Since the total available frequency spectrum is limited, we are interested in the maximum data rate that can be transmitted over a channel with limited bandwidth and an existing noise and interference level. In principle the data rate (in bps or bit/s) can be increased by increasing the number of bits per time. For simple modulation schemes, such as binary phase shift keying (BPSK see detailed explanation in Chap. 4), more bits/s consequently will increase the frequency spectrum of the signal and thus will require a larger bandwidth. Other modulation schemes such as higher order QAM (see Chap. 4) put the information bits, into defined phase shift and also into amplitudes. Obviously these systems require lower noise and interference within the channel, so that the receiver can reconstruct the original transmitted signal with its fine amplitude steps. Thus a higher data rate can be obtained, with a given signal power, with lower noise and interference, that is a higher signal-to-noise ratio. The possible maximum data rate is called the channel capacity. Claude Shannon published the first channel capacity estimation for digital data transmission stating that the total maximum channel capacity C, or maximum data rate, depends basically on the available bandwidth B and the signal-to-noise ratio SNR = S/N with S as signal power and N as noise power level [14].

$$\mathbf{C} = \mathbf{B} \cdot \mathbf{log}_2\left(1 + \frac{\mathbf{S}}{\mathbf{N}}\right). \tag{2.5}$$

This formula represents the maximum limit of the channel capacity C depending on the available spectrum or bandwidth B and the signal-to-noise ratio SNR. To be more precise, the SNR normally considers the ratio of the signal power level to all disturbing factors like noise and interferences. In the general case of wireline or wireless transmissions, not only the distance-dependent reduction of signal power S during the wave propagation in the air, the thermal noises N of transmitter and receiver circuits, as well as the transmission medium, but also the interferences I from the neighboring transmitters in the same frequency band, will reduce the signal-to-noise ratio. In some publications this term is also generally written as SINR = S/(I+N), so that the Shannon capacity formula can be written also as [5]

$$\mathbf{C} = \mathbf{B} \cdot \mathbf{log}_2\left(1 + \frac{\mathbf{S}}{\mathbf{I} + \mathbf{N}}\right). \tag{2.6}$$

Even though Shannon did not explicitly explain how to achieve the best channel capacity, this formula is the best approach to estimate the maximum channel capacity limit. In fact, all researchers working towards the goal to achieve the maximum capacity, to approach this maximum channel capacity limit, use sophisticated coding and modulation schemes as well as forward error correction techniques to exploit the given channel conditions.

In free space optical communications and deep space optical communications, the channel conditions become extremely critical, so that the SINR becomes extremely small and the total achievable bitrates will be strongly reduced. Whereas the optical fiber is definitely the perfect transmission medium with high SINR enabling extremely high bitrates up to Tbps by using higher order modulation schemes 4PSK/QPSK, 16-QAM, 64-QAM, 128-QAM and 256-QAM, free space optical communications or even deep space optical communications can only use robust modulation schemes like M-PPM in order to overcome the extremely high path loss and in order to achieve acceptable, relatively low bit rates. These modulations techniques will be discussed in the next chapter.

Higher order modulation schemes, can achieve higher data rates like 16-QAM, 64-QAM, 128-QAM, and 256-QAM, or can guarantee robust, errorfree transmission by using more time slots for one symbol like M-PPM which contains a certain number of bits. PPM modulation schemes are robust, but less spectral efficient, since a large number of time slots represent fewer bits in comparison with M-QAM modulation schemes. In order to compare different modulation schemes with the same criteria, for example in case of bit error rate (BER) calculation, normally the energy per bit to noise spectral density (E_b/N_0) will be used:

$$\frac{E_b}{N_0} = \frac{S}{I+N} \cdot \frac{W}{R} \tag{2.7}$$

with W as bandwidth and R as data rate. E_b/N_0 can be considered as normalized value of energy per symbol E_s/N_0 with ρ as spectral efficiency.

$$\frac{E_b}{N_0} = \frac{1}{\rho} \cdot \frac{E_s}{N_0}. \tag{2.8}$$

2.3 Free Space Optical Link Model

The block diagram of a FSOC system contains three primary subsystems: transmitter, free space optical communication channel, and receiver [11].

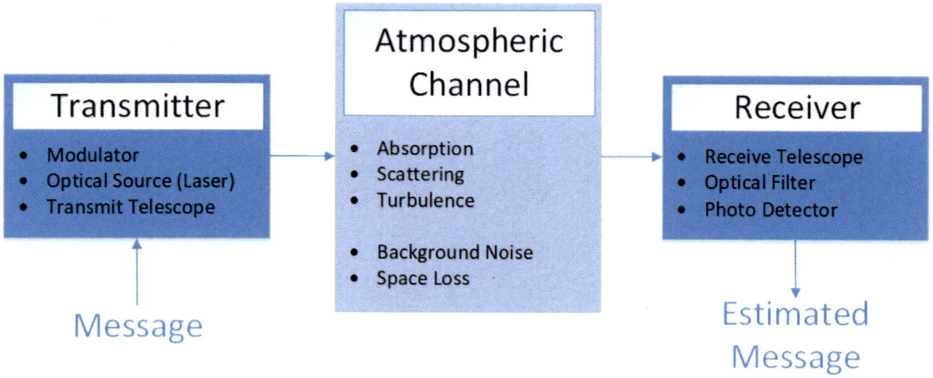

Fig. 2.1 Schematic model of a free space optical communication system

2.3.1 Transmitter

Its primary function is to modulate the incoming message signal onto the optical carrier which is then propagated through the atmosphere to the receiver. Essential components are the modulator, the optical source which is most likely a laser diode, and a transmitting telescope or optical antenna. The telescope is not only responsible for beam shaping but also the movable part for beam tracking (Fig. 2.1). Most famous transmitters used in fiber optical communications are the DFB and DBR laser diodes (Fig. 2.2) (see for example [5]), which will be partially used in free space optical communications, if necessary in combination with an EDFA (Erbium-doped fiber amplifier). More details will be discussed in Chap. 4.

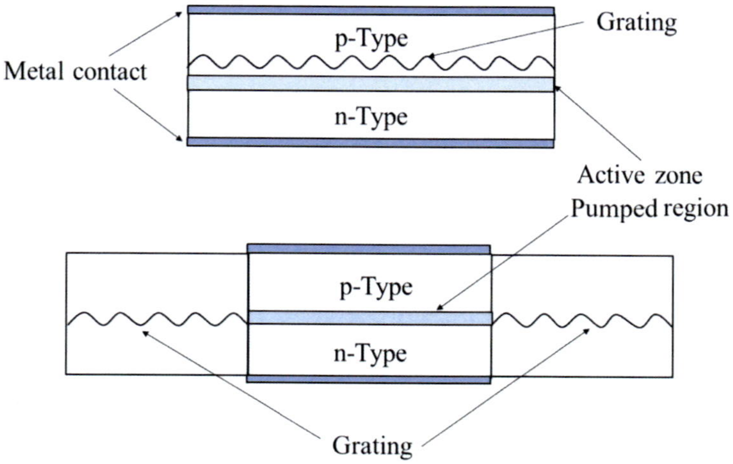

Fig. 2.2 Distributed Feedback (DFB) and Distributed Bragg Reflexion (DBR) laser diode mostly used in the fiber optical communications [5]

2.3.2 Receiver

A receiver is used to recover the transmitted data. It consists of a receiver telescope which is the equivalent to the transmitter telescope, a very sharp optical bandpass filter which reduces the background noise and limit the received signal bandwidth, and the demodulator. The APD (Avalanche Photodiode) (Fig. 2.3) is the most famous receiver used in fiber optical communications, which will be partially used in free space optical communications. In extremely deep space optical communication scenarios, if only few photons are received, also Superconducting Single Photon Counters SSPC will be used. More details will be discussed in Chap. 4.

2.3.3 Atmospheric Channel

The atmospheric channel is a major limiting factor of FSOC since unpredictable conditions like snow, fog, rain, or clouds can influence the strength of the received signal. It is important to mention that these factors do not have fixed characteristics and change significantly with time.

The Earth is surrounded by a layer of molecules or gases known as the atmosphere, which extends hundreds of kilometers above the surface. Nitrogen and oxygen make up over 98% of this mixture, with about 78% nitrogen constituting and around 21% oxygen [17]. The atmosphere is classified into four major layers, primarily distinguished by how temperature changes with altitude. These layers are separated by three distinct isothermal boundaries where temperature remains nearly constant (Fig. 2.4). These layers are as follows [13, 18]:

Troposphere: This lowest layer reaches up to approximately 11 km and contains about 75% of the atmosphere's total mass. Temperature is highest near the Earth's surface and

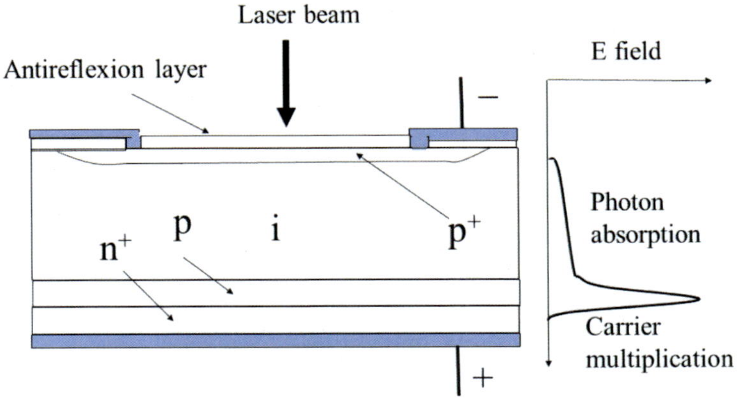

Fig. 2.3 Avalanche Photodiodes (APD), also used in the fiber optical communications [5]

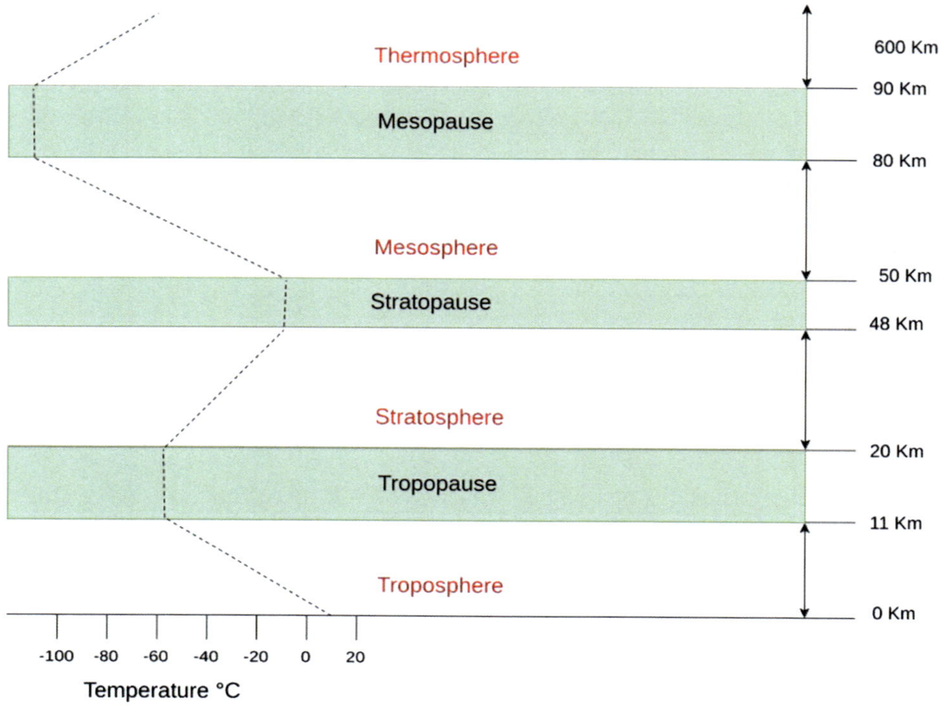

Fig. 2.4 Atmospheric layers with altitude, adapted from [19]

decreases steadily with height, reaching around −55°C near the top. Above the troposphere lies the tropopause, an isothermal region between 11 and 20 km where the temperature remains stable at some −55°C. Together they form lower atmosphere.

Stratosphere: Extending from about 20–48 km, this layer experiences a temperature increases with altitude, mainly due to ozone molecules absorbing ultraviolet radiation from the sun. The ozone layer, which also protects living organisms by filtering harmful UV rays, is concentrated between roughly 10 and 50 km. The stratopause, located between 48 and 50 km, is another isothermal boundary separating the stratosphere from the next layer.

Mesosphere: From around 50–80 km, temperatures decline steadily with altitude, reaching the coldest point in the atmosphere at approximately −90°C near 90 km. The mesopause, between 80 and 90 km, is the final isothermal boundary. The stratosphere and mesosphere together with their respective boundaries are often referred to as the middle atmosphere.

Thermosphere: This uppermost layer extends from the mesopause up to around 600 km or more. Temperatures increase sharply with altitude due to the absorption of solar energy. The thermosphere contains most of the ionosphere, a region highly ionized by solar UV radiation starting around 70 to 80 km and extending to over 1000 km. The ionosphere plays an essential role in radio wave propagation. The thermosphere also includes the exosphere, which is the outermost region of Earth's atmosphere.

The selection of optical wavelengths depends on multiple factors. For terrestrial FSOC systems costs, eye safety, and the transmission windows are critical. Regarding eye safety regulations, 1.5 μm and above wavelengths are always the preferred option since wavelengths between 400 and 1400 nm are hazardous because they are focused directly within the human's eye on the retina [20]. The transmission windows arise from the radiation absorption of molecules in the atmosphere. Certain gases such as water, carbon dioxide or ozone absorb electromagnetic waves at a particular frequency. In contrast to these absorption bands, there are spectral areas with little or no absorption to specific wavelengths, these are known as atmospheric windows. These wavelengths are preferred for a low attenuation optical transmission [21]. Other limiting factors like scattering are also dependent on the wavelength. This is described more in detail in Chap. 3.

2.4 Modulation Schemes OOK, BPSK, QPSK, M-QAM, and M-PPM

Different Modulation schemes are described in many books and papers [25–27] and will be discussed in the following sections. These are On-Off Keying OOK, Binary Phase Shift Keying BPSK, M-Quadrature Amplitude Modulation M-QAM, and Pulse Position Modulation PPM. These schemes will also be explained in the so-called constellation diagram. This constellation diagram depicts the different modulation states of the signal in the so-called IQ phase plane. In phase (I) represents 0° phase shift (horizontal line), whereas quadrature phase (Q) represents a 90° phase shift (vertical axis). It allows to see the respective phase and amplitude of a digital signal. The amplitude is represented by the distance of a point to the origin.

2.4.1 On-Off Keying (OOK)

On-Off Keying (OOK) modulation (Fig. 2.5) is a very simple modulation technique in which binary data is represented by the presence or absence of a single light pulse in each symbol interval T. The binary information sequence can be directly mapped to a sequence of light pulses at the transmitter using the following rule: If the information bit is 1, a laser pulse will be transmitted. If the information bit is 0, nothing will be transmitted. By using OOK, the phase is irrelevant, only the amplitude or power are important. Either the transmitter power is 1 or 0. In order to compare OOK with 2-ASK and BPSK, the off state corresponds to I = 0, Q = 0, whereas the on-state corresponds to I = 1, Q = 0. In other words, the amplitude for on-state will be 1, phase is assumed to be 0.

One light pulse thus corresponds to one bit. Depending on the transmission power, very long distances can be achieved with this technique. However, it is very susceptible to noise

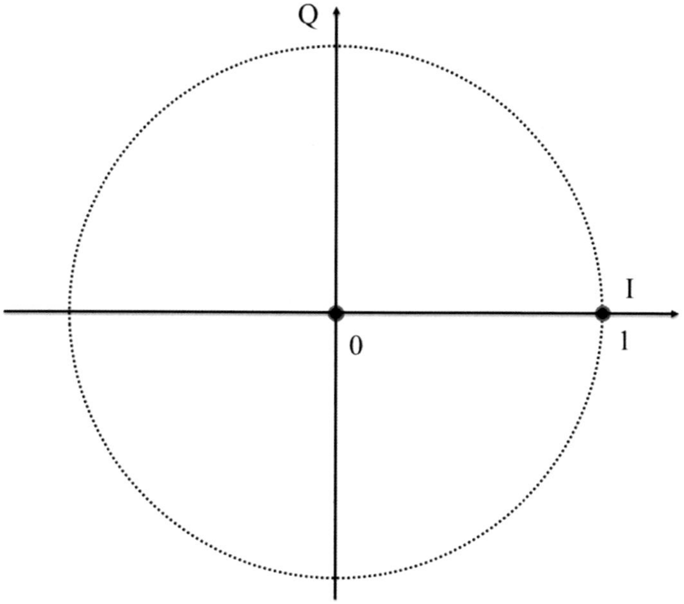

Fig. 2.5 On-Off-Keying (OOK) modulation scheme. For example 0 corresponds to I = 0, Q = 0; 1 corresponds to I = 1, Q = 0

and has a limited data rate; therefore, other modulation techniques are used for very long distances [9, 12].

2.4.2 BPSK, QPSK, DP-QPSK

Differently to the On-Off-Keying technique where the laser pulse will be transmitted (1) or not (0), BPSK (Binary Phase Shift Keying) uses a unique light pulse for 1 and also for 0, but with a phase shift of 180° (Fig. 2.6).

Similarly QPSK (Quadrature Phase Shift Keying) uses a 90° phase shift between the 4 constellation states which represent now the combination of 2 bits (00, 01, 10, 11), so that QPSK transmits two bits per symbol (Fig. 2.7). By utilizing both polarization planes to further increase the efficiency, if the channel condition allows, DP-QPSK can thus transmit four bits per symbol. This modulation method is also only used over short transmission distances and provides similar data rates to 16-QAM [11].

Dual Polarization-Quadrature Phase Shift Keying (DP-QPSK) combines quadrature phase shift keying (QPSK) with dual polarization, i.e. utilizing two orthogonal polarization planes of light, each with QPSK, to double the data rate.

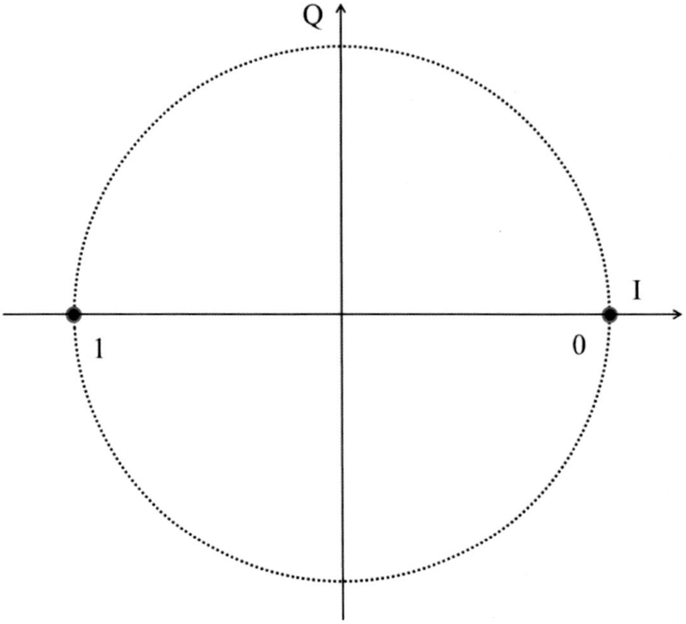

Fig. 2.6 Binary Phase Shift Keying (BPSK) modulation scheme

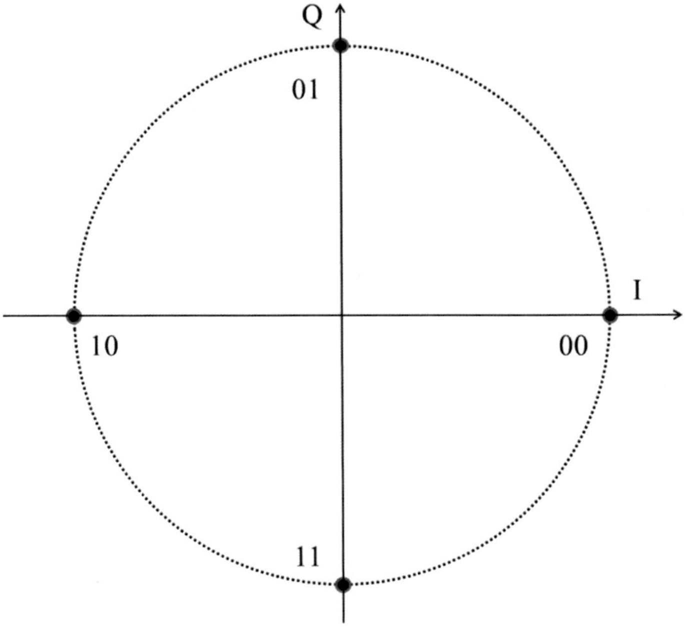

Fig. 2.7 Quadrature Phase Shift Keying (QPSK) modulation scheme

2.4.3 16-QAM, M-QAM

16-Quadrature Amplitude Modulation (16-QAM) is a modulation technique used not only in radio but also in optical communications (Fig. 2.8). It combines amplitude and phase modulation to transmit more information per symbol. In 16-QAM, several amplitude and phase levels are used, resulting in a total of 16 different states. This increases the data rate compared to simpler modulation methods. However, 16-QAM requires higher signal quality, i.e. SINR, and more precise synchronization, making it more susceptible to noise and interference, but also more efficient in bandwidth utilization. Therefore, this modulation type is only used for short distances. Possible data rates here, for example, with a transmit aperture of 10 cm, were 10 Gbit/s up to 60 km or even 100 Gbit/s over a distance of over 20 km [11].

By further increasing the number of possible constellation states or number of the M different shifted amplitudes and phases, higher bit rates can be achieved, as long as the channel conditions permit, for example 64-QAM in (Fig. 2.9). In an optical fiber 256-QAM can be used to achieve about 1 Tbps data rate. The higher the modulation order M, the higher the bit rates can then be achieved.

The Bit Error Probability of QAM modulation schemes depends on the energy per bit to spectral noise density (Eb/No) or SINR.

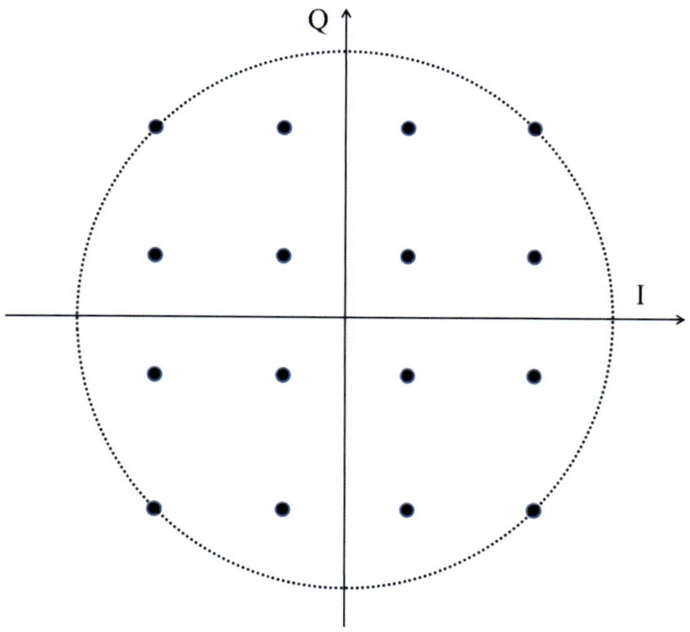

Fig. 2.8 16-QAM modulation scheme

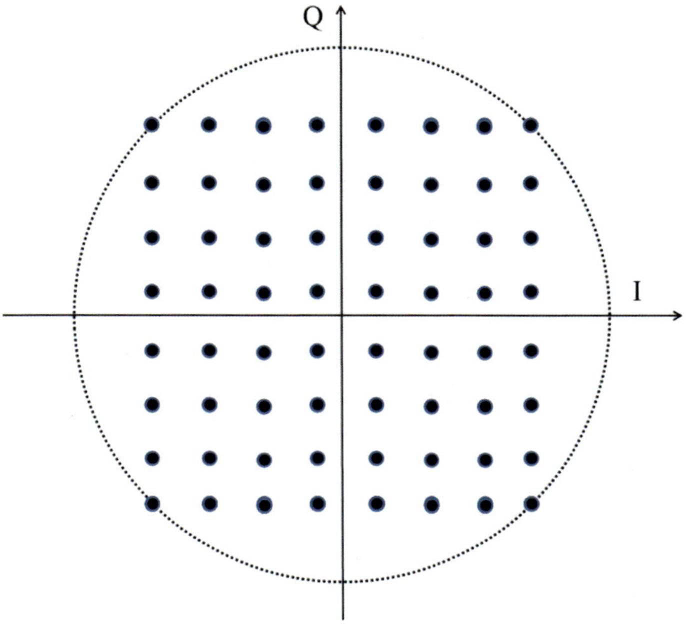

Fig. 2.9 64-QAM modulation scheme

The bit error rate of for example 1% for acceptable data transmission by using M-QAM modulation schemes requires a very high SINR (signal-to-noise ratio) or Eb/No (ratio of energy per bit to spectral noise density) to be higher than 10–20, cf. Figure 2.10. This condition can be guaranteed only by using optical fiber communication technologies with extremely low attenuation (0.24 dB/km) and low noise and interference levels in the transmitter and receiver systems.

In 2016, Nokia Bell Labs, Deutsche Telekom T-Labs, and the Technical University of Munich (TUM) have achieved unprecedented transmission capacity and spectral efficiency in an optical communications field trial with a new modulation technique known as Probabilistic Constellation Shaping (PCS) (see for example [28]) which in principle modifies quadrature amplitude modulation (QAM) formats to achieve a world record transmission capacity of 1 Tbps over a given channel to significantly improve the spectral efficiency of optical communications and almost approach the Shannon Channel Capacity limit. Traditionally, all QAM constellation points have the same amplitudes. PCS cleverly uses constellation points with high amplitude less frequently than those with lower amplitude to transmit signals that, on average, are more resilient to noise and other impairments. This allows the transmission rate to be tailored to ideally fit the transmission channel, delivering up to 30 percent greater reach. This is only possible, because the optical fiber provides the perfect transmission channel quality, so that the up to 256-QAM can be applied.

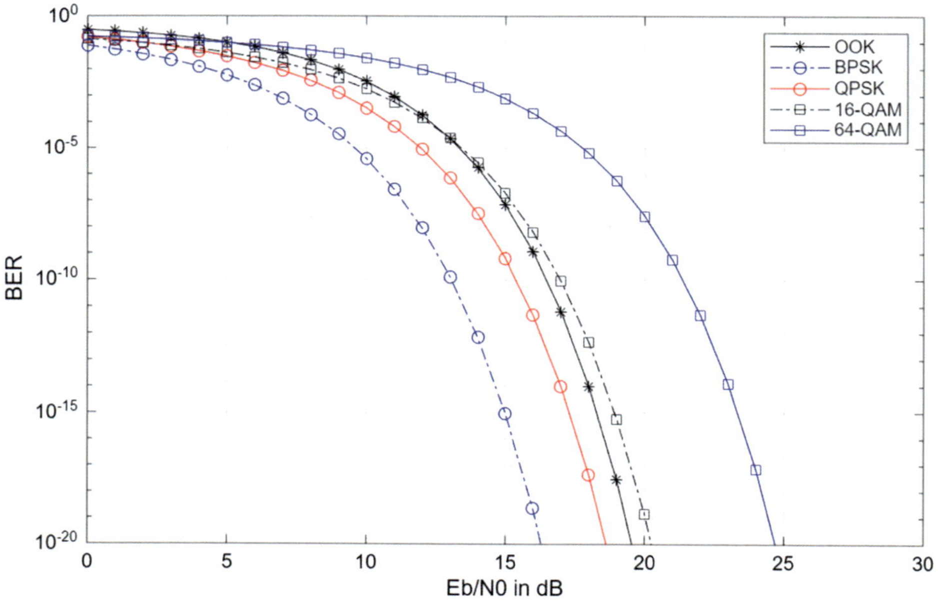

Fig. 2.10 Bit Error Probablity of the modulation schemes OOK, BPSK, QPSK, 16-QAM, and 64-QAM depending on the Eb/No or proportionally on the signal-to-noise ratio SNR

In free-space optical communications only a very short distance, in combination with appropriate laser diodes and photodiodes, will enable such a signal-to-noise ratio in order to achieve the higher bit rates. In our projects we investigated the modulation schemes up to 16-QAM used for short distance FSOC [11].

It should be mentioned, that all these modulation schemes that use the phase information require a reference phase which means a phase locked laser at the detector side. This phase locking requires a reference laser with low phase noise and a signal with relatively low noise and interference.

In deep space optical communications with extremely long distances between the transmitter and the receiver such "perfect" signal-to-noise ratios can never be achieved. Therefore other less efficient modulation schemes, with respect to bit rates, but most robust modulation schemes will be the indispensable solution, in order to receive the data with low bit rates over extremely long distance.

2.4.4 PPM in Comparison with the Classical OOK

In free space optical communications, if the distance between the transmitter and receiver is relatively short, i.e. significantly less than 100 km, and the weather condition is ideal, then QPSK can be used. For shorter distance even 16-QAM can be applied.

In case of deep space optical communications, the total transmission link budget is extremely poor due to the long distance and the atmospheric disturbances, even On-Off-Keying (illustrated in Fig. 2.12) cannot be used. A less efficient, in the terms of data rates, but much robust modulation technique Pulse Position Modulation (PPM) is used (Fig. 2.11).

In PPM a specific time interval is divided into different time slots. The message in form of bits is not directly modulated as in the case of On-Off-keying, but rather the individual timeslot positions, where the PPM pulse is allocated, defines the bit information, also depending on the PPM order.

It can be recognized clearly in Fig. 2.11, that the higher the PPM order M, the more robust the M-PPM will be, i.e. the BER decreases with increasing PPM order M for the same Eb/No. That is also the reason why PPM is used for extremely poor Eb/No situation which is the case of DSOC.

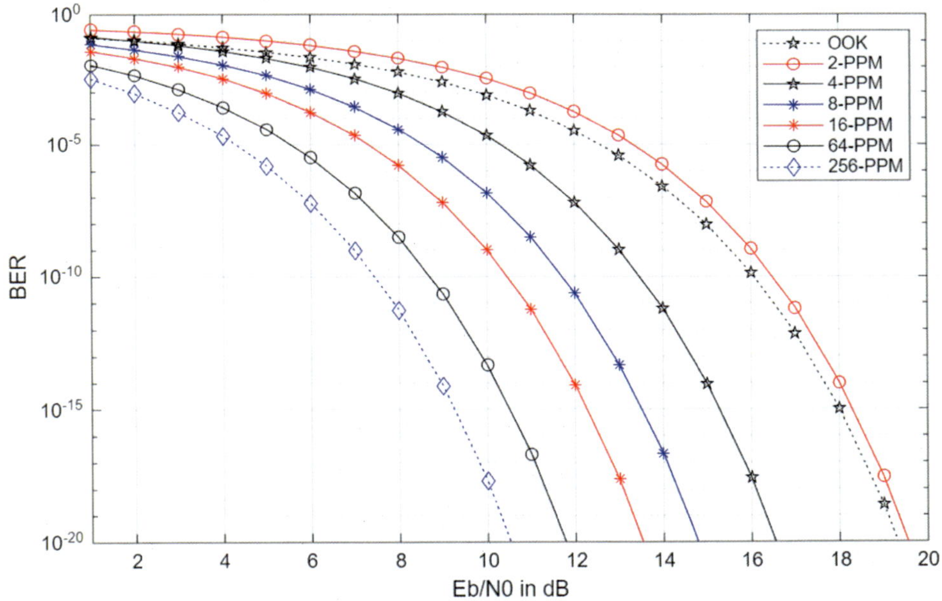

Fig. 2.11 Bit Error Probablity of the M-PPM modulation schemes depending on the Eb/No or proportionally on the signal-to-noise ratio SNR

Fig. 2.12 Classical On-Off-Keying modulation scheme

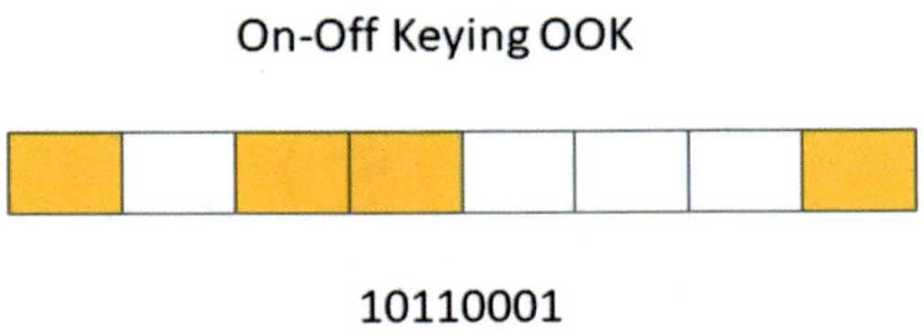

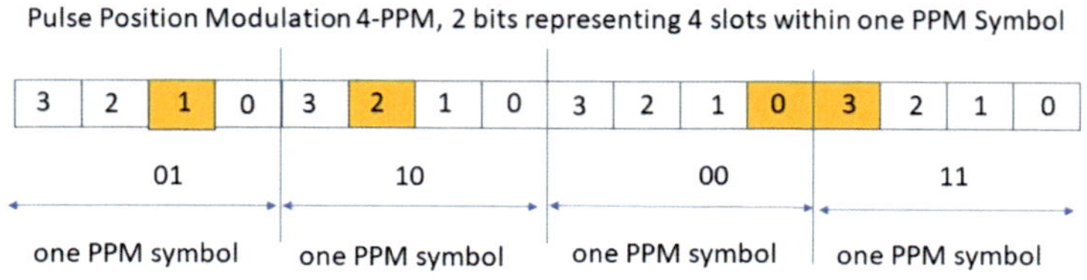

Fig. 2.13 4-PPM modulation scheme

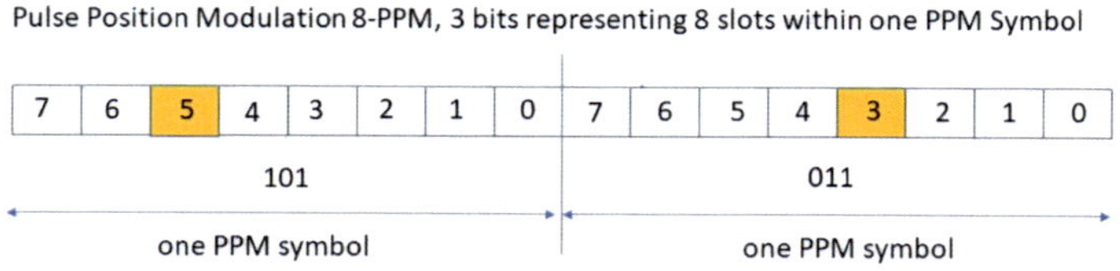

Fig. 2.14 8-PPM modulation scheme

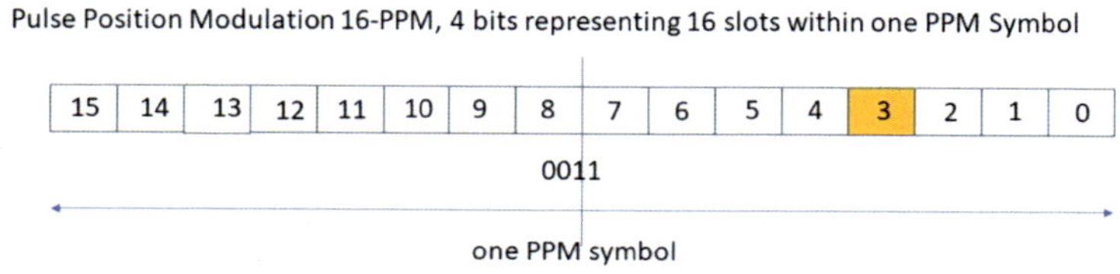

Fig. 2.15 16-PPM modulation scheme

For example, the time interval is divided into four slots, that is M=4. Each slot corresponds to a value between zero and three, i.e., a total of two bits (Fig. 2.13). By sending one pulse per unit of time in the correct slot, the message can then be modulated (Fig. 2.15).

In case of OOK, each slot represents one bit.

In case of 4-PPM, one symbol contains 4 slots, representing 2 bits (Fig. 2.13).

In case of 8-PPM, one symbol contains 8 slots, representing 3 bits (Fig. 2.14).

In case of 16-PPM, one symbol contains 16 slots, representing 4 bits.

The final achievable bit rates depend not only on the slot period, but on the additional slots for channel coding and Forward Error Correction FEC. The latter slots are omitted in this book, in order to show only the principle and ideas of the PPM. This means that by considering the channel coding and FEC, the payload data rates will be reduced correspondingly.

PPM requires low average power, achieves reasonably information efficiencies and is resistant to background noise, making it particularly suitable for both FSOC and DSOC [9, 10]. The lower the signal-to-noise ratio, the higher the modulation order M is normally chosen. The modulation order M can only be increased to a certain limit (see for example [10]), since the slot period cannot be chosen arbitrarily small, to avoid the overlapping slots leading to bit detection errors.

Table 2.3 Comparison of the maximum spectral efficiency of the different modulation schemes.

Modulation scheme	Spectral efficiency in bit/s/Hz
64-PPM	0.094
32-PPM	0.156
16-PPM	0.25
8-PPM	0.375
4-PPM	0.5
BPSK	1
QPSK	2
8PSK	3
16PSK/16-QAM	4
64-QAM	6
256-QAM	8

2.4.5 Spectral Efficiencies of the Modulation Schemes M-QAM and M-PPM

The maximum spectral efficiencies of different modulation schemes are compared with each other (Table 2.3). The maximum spectral efficiency corresponds to transmission spectral efficiency of all transmitted symbols and intervals, including not only the payload data, but also the coding rate by considering the intervals for forward error correction (FEC) or channel coding, guard intervals, and synchronization overhead. Correspondingly the payload data rates will be reduced. One can clearly recognize that with M-QAM under good or perfect channel conditions, i.e. SINR in best case for fiber optical transmission, higher spectral efficiencies can be achieved.

In case of critical channel conditions like deep space optical communications (DSOC), the SINR is extremely low and only a few photons can be received. Therefore only the very robust, but less spectral efficient M-PPM can be used to overcome the extremely long wave propagation distance.

2.5 Beam Divergence Loss

The most significant loss in FSOC is usually the beam divergence loss which describes the attenuation between two antennas or telescopes (Fig. 2.16) [12].

The divergence loss is dependent on the propagating beam width and the diameter of the receiving telescope which can be regarded as an equivalent to the antenna aperture in radio

Fig. 2.16 Schematic description of beam divergence

transmission. It also depends on the propagating wavelength because of the inclusion of the free space loss. The formula for the calculation of the received power is:

$$P_R = P_T \cdot G_T \cdot G_R \cdot L_P \tag{2.9}$$

where L_P is the free space path loss, G_T and G_R are the effective gains of receiver and transmitter, P_T is the transmitted power and P_R the received power. The gain of both optical antennas or telescopes is approximated by:

$$G = \frac{4\pi A}{\lambda^2} \tag{2.10}$$

where λ is the wavelength of the transmitted signal and A the effective optical antenna area.

Depending on the efficiency and the optical telescope characteristics, the total gain can vary. With this equation in mind, a narrow beam or low divergence is preferable with the condition that the antennas are perfectly aligned. Otherwise active beam tracking and pointing systems are required to reduce the pointing loss that occurs when the center of the beam does not directly faces the receiver. The free space loss factor is given by:

$$L_P = (\frac{\lambda}{4\pi R})^2 \tag{2.11}$$

where R is the distance between both optical antennas.

2.6 Influence of Changing Weather Conditions, Turbulence, and Scintillation

As already mentioned in the introduction, the quality of an FSOC link can vary strongly depending on weather conditions that affect the line of sight. Mostly fog, rain, and snow decrease the received signal power. Additionally, attenuation due to particles must be considered. If the number of particles in the atmosphere reaches a certain threshold, a complete link outage can occur. For a theoretical approach, this effect is divided into three different ranges regarding the size of the particles related to the wavelength. Fog consists of tiny particles that lead to Mie-scattering, in contrast raindrops and snowflakes are comparatively larger and produce deeper fades in the signal [22, 23]. The effect of fog can be calculated with the Mie-scattering theory which is very sophisticated and requires detailed fog parameters. Alternatively, an approach based on the visibility range information can be used [24]. It defines the attenuation coefficient of fog given by an empirical model for Mie-scattering. The specific attenuation in dB/km is derived by

$$A(\frac{\text{dB}}{\text{km}}) = 10 \cdot \log(\gamma(\lambda)) \tag{2.12}$$

$$\gamma(\lambda) = \frac{3.91}{V}(\frac{\lambda}{550\,\text{nm}})^{-p} \tag{2.13}$$

where V stands for the visibility range and λ is the operating wavelength. p is the size distribution coefficient that describes the "thickness" of the fog. 550 nm is the reference wavelength used in this approach. The Kruse model [9] defines p for different visibility factors:

$$p = \begin{cases} 1.6 & V > 50 \\ 1.3 & 6 < V < 50 \\ 0.585 \cdot V^{1/3} & V < 6 \end{cases} \tag{2.14}$$

Unaffected by the visibility range, the attenuation factor for fog is always dependent on propagating wavelength. A larger wavelength leads to lower attenuation (Fig. 2.17). Rain and snow attenuation are not affected by the propagating wavelength, as these particles are much larger than wavelength. The specific attenuation increases with the rainfall or snowfall rate.

Inhomogeneous temperature and pressure lead to various refractive index values in the atmosphere that directly influence the transmitted laser beam negatively. Turbulent cells are formed by these factors. These cells, also called eddies, have various sizes and different refractive indexes. Because these cells are distributed very randomly, calculating this influence is very sophisticated. When an optical signal is distributed through a turbulent atmosphere, it causes wave front distortion in phase and amplitude. This can lead to a complete link failure. When the cells are larger than the beam size, a phenomenon called beam

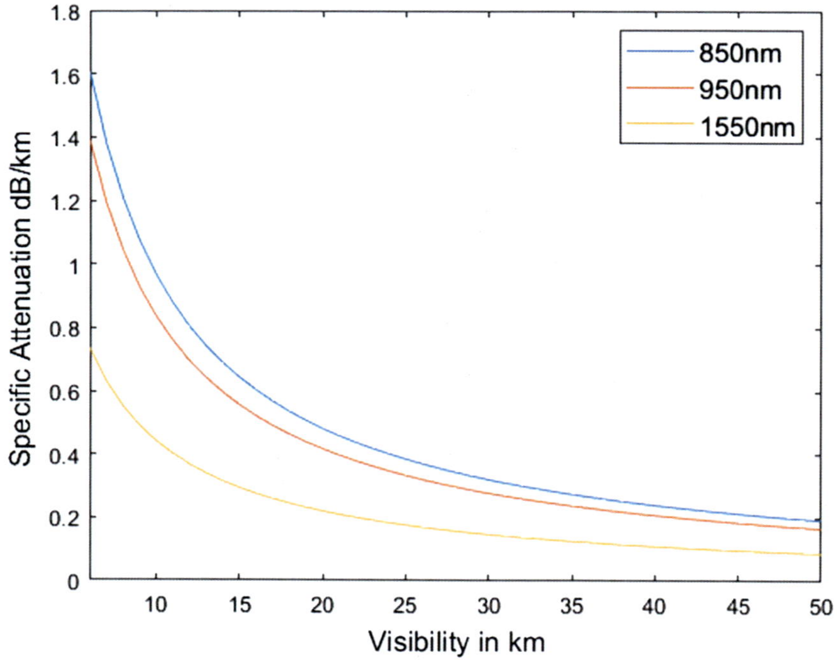

Fig. 2.17 Specific attenuation versus distance for fog depending on wavelength

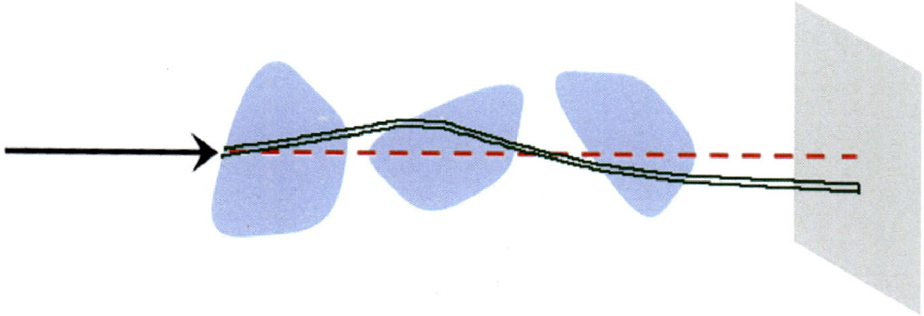

Fig. 2.18 Influence of turbulence cells on propagation angle

wandering occurs. Due to the change of refractive indexes in the cells, the distribution angle of the beam changes (Fig. 2.18).

References

1. R. E. Collin, F. J. Zucker: Antenna Theory, Part 1. McGraw-Hill Book Company (1969).
2. H.-G. Unger: Planar Optical Waveguides and Fibres. Oxford Clarendon Press (1977).
3. C. A. Balanis: Antenna Theory. John Wiley & Sons, Inc. Fourth Edition (2016).
4. K. W. Kark: Antennas and Radiation Fields (in German: Antennen und Strahlungsfelder). Springer Vieweg (2018).
5. S.-P. Chen: Fundamentals of Information and Communication Technologies. Cambridge Scholars Publishing (2020).
6. S.-P. Chen, H. Schmiedel: RF Antenna Beam Forming. Focusing and Steering in Near and Far Field. Springer Nature (2023).
7. S.-P. Chen, J. Villalvilla: Comparison of Modified Woo's Solar Phase Scintillation Model with ESA's BepiColombo Superior Solar Conjunction Measurement Data for X-Band and Ka-Band. CEAS Space Journal, Springer Nature (2022).
8. S.-P. Chen: Investigations of free space and deep space optical communication scenarios. CEAS Space Journal, 14, 357–364. https://doi.org/10.1007/s12567-021-00388-y Springer Nature (2021).
9. H. Hemmati: Deep Space Optical Communications. Wiley, 2005.
10. Aboagye, E. D., Chen, S.-P. (2021). Deep space optical communications (DSOC) downlink simulation with varying PPM order. Optical and Quantum Electronics, 53, 583. https://doi.org/10.1007/s11082-021-03232-z
11. Tobias Siegel, Shun-Ping Chen: Investigations of Free Space Optical Communications unter Real-World Atmospheric Conditions. Wireless Personal Communications, Springer Nature. https://doi.org/10.1007/s11277-020-07724-1, August 12, 2020.
12. F. Schüßler, T. Stieglitz, P. Thielke, S.-P. Chen (Supervisor): Absorption and Scintillation of Free Space Optical Communications. IWP Team Research Project (in German). Faculty of Electrical Engineering and Information Technology, h-da Darmstadt University of Applied Sciences. June 19, 2024.
13. R. G. R. Charles, S. Sukumaran, V. Jose, V. Sadashivpet, S.-P. Chen (Supervisor): FSOC (Free Space Optical Communications) und DSOC (Deep Space Optical Communications). Team Research Project. Faculty of Electrical Engineering and Information Technology, h-da Darmstadt University of Applied Sciences. September 16, 2025.
14. C. Shannon: C. Shannon, "A Mathematical Theory of Communication," The Bell System Technical Journal (1948).
15. Matlab. https://www.mathworks.com/products/matlab.html.
16. OptiSystem. Simulation system of OptiWave Inc. https://optiwave.com/optisystem-overview/.
17. L. C. Andrews, R. L. Phillips, and C. Y. Hopen, Laser Beam Scintillation with Applications, SPIE Press, Bellingham, Washington, 2001. https://doi.org/10.1117/12.3.412858.
18. Alan Buis and the NASA Science Editorial Team, Earth's Atmosphere: A Multi-layered Cake, Available: https://science.nasa.gov/earth/earth-atmosphere/earths-atmosphere-a-multi-layered-cake/, October 2, 2019. Accessed: 2025-08-17.
19. Larry C. Andrews and Melissa K. Beason, Laser Beam Propagation in Random Media: New and Advanced Topics, SPIE Press, 2023. Available: https://spie.org/Publications/Book/2643989.
20. R. Long: Atmospheric attenuation of ruby lasers, IEEE, 1963. https://doi.org/10.1109/PROC.1963.2294.
21. A. S. Jursa: Handbook of Geophysics and the Space Environment. Air Force Geophysics Laboratory Air Force Systems Command of the United States Air Force. 1985.
22. M. Rouissat, R . Borsali, M. Chick-Bled: Free Space Optical Channel Characterization and Modeling with Focus on Algeria Weather Conditions. International Journal of Computer Network and Information Security, Vol. 4, Pages. 17–23, 2012.

23. M. C. Alnaboulsi, F. de Fornel: Fog attenuation prediction for optical and infrared waves. Optical Engineering, 2004. https://doi.org/10.1117/1.1637611.
24. I. I. Kim, B. McArthur, E. J. Korevaar: Comparison of laser beam propagation at 785 nm and 1550 nm in fog and haze for optical wireless communications. SPIE Journal, 2001.
25. B. Sklar: Digital Communications: Fundamentals and Applications. Second Edition. Prentice Hall P T R, 2001.
26. M. B. Sklar: Introduction to Digital Communications. Pearson Prentice Hall, Pearson Educations, Inc. 2005.
27. X. Shi, Y. Sun, J. Tian, M. Chen, Y. Liu, N. Xie, J. Zhang: BER Performance Analysis of Non-Coherent Q-Ary Pulse Position Modulation Receivers on AWGN Channel. MDPI Sensors 2021, 21, 6102. https://doi.org/10.3390/s21186102.
28. Optical fiber transmits one terabit per second. https://www.tum.de/en/news-and-events/all-news/press-releases/details/33399

Atmospheric Absorption and Scintillation 3

The link performance of Free Space Optical Communications (FSOC) and Deep Space Optical Communications (DSOC) is investigated by considering two scenarios in space communications. Firstly, for downlinks and uplinks between Earth ground stations and near Earth geostationary (GEO) satellites, and secondly between the Earth and a spacecraft with a large distance of 1 astronomical unit (AU) to Earth or more. Generally a distance larger than 0.01 AU or approximately 1,500,000 km from Earth is considered as deep space. In these theoretical investigations, different realistic system parameters for the optical lasers as transmitters, avalanche photodiodes as receivers (APDs), optical telescopes, atmospheric disturbances like scintillation, and absorption are considered. Simulations have been conducted and will be explained in the following chapters. The simulation results are compared with existing project data and valuable ESA experimental results in order to verify and improve the simulation models. The comparison in this book shows that the simulation models for the link budget and the scintillation estimation are feasible for the investigations of FSOC and DSOC, and can be used to investigate improved design and implementation of DSOC projects for planned long-term and medium-term space missions.

3.1 Introduction

Laser optical communication links show significant advantages in comparison with classical radio frequency microwave links for space communication technologies, in terms of link performance, system size, and achievable transmission bitrates. Many projects have been proposed by the member agencies of the International Astronautical Federation (IAF) like NASA, ESA etc. and trials have been successfully carried out in the last decades. The link performance of FSOC and DSOC is investigated by considering different scenarios in space

S.-P. Chen, *Fundamentals of Free Space and Deep Space Optical Communications*,
https://doi.org/10.1007/978-3-032-35826-4_3

communications. For example, downlink and uplink between Earth ground stations and a) satellites in low Earth orbit (LEO) or geostationary (GEO) satellites, or b) spacecrafts in the lunar orbit, or c) spacecrafts with large distances of about 1–3 AU to Earth. In these theoretical investigations, different realistic system parameters for the optical transmitters, receivers, optical telescopes, and atmospheric disturbances like scintillation and absorption are considered. For extremely large distances the robust pulse position modulation scheme (PPM) and APD photon counting receivers help to achieve the desired transmission bitrates. Various results like transmitter and receiver power densities, laser beam spot sizes, and received power at the receiver are simulated. The results are compared with some existing project planning data and experimental results in order to verify and improve the simulation model. The investigation results can be utilized in the design and implementation of the DSOC projects for planned long-term and medium-term space missions.

3.2 Theoretical Background

Compared with classical radio frequency deep space communications in S, X, and Ka band, deep space optical communications show advantages in terms of beam divergence (improvement by a factor of 100), high bitrate delivery with significantly reduced aperture size of the flight terminal.

The demonstration and measurement projects NASA LLCD (Lunar Laser Communication Demonstration), ESA ARTEMIS SILEX (Semiconductor-laser Inter-satellite Link EXperiment) and ARTEMIS to OGS (Opical Ground Station) have been carried out, also with cooperation between NASA and ESA [1–4]. Many suggestions concerning the efficient transmitter telescope, APD photon counting receiver system [5], and PPM modulation scheme have been proposed to achieve the optimum system performance due to extremely large distances in the case of deep space optical communications. Also future projects are being planned like PSYCHE mission and DOCS L5 [6, 7].

For the general link budget calculation, in free space optical communications, plane wave approximation, spherical wave propagation, and Gaussian beam propagation models are commonly used [8, 9]. We focus on spherical wave propagation and Gaussian beam propagation. These two methods will deliver identical results, if the same planning parameters are used.

The free space optical channel model can be schematically described by Fig. 3.1.

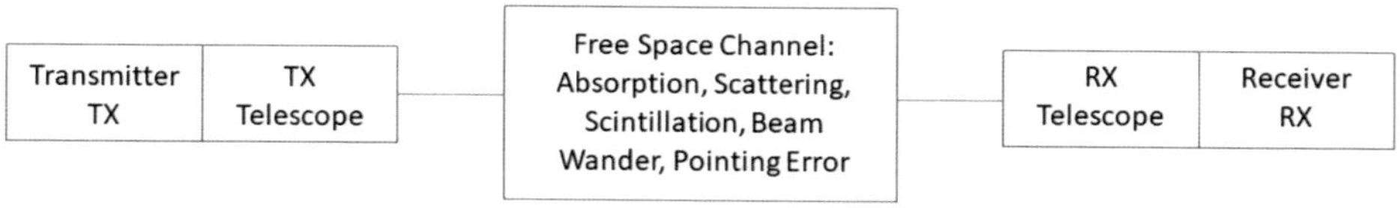

Fig. 3.1 Free space optical communications (FSOC) channel model

3.2.1 Spherical Beam Propagation

The link budget calculation of the spherical wave propagation is

$$P_R(L) = P_T \cdot \eta_T \cdot g_T \cdot \eta_{Atm}(L) \cdot \eta_{FS}(L) \cdot g_R \cdot \eta_P \cdot \eta_R \tag{3.1}$$

with P_T, $P_R(L)$ as transmitted power and received power at a distance L; η_T, η_R as transmitter and receiver telescope aperture efficiencies; η_P as pointing efficiency; g_T, g_R as transmitter and receiver aperture gain; $\eta_{Atm}(L)$ as the atmospheric transmittance due to absorption loss, Rayleigh scattering (size of the scatterers $\ll \lambda$) and Mie scattering (size of the scatterers $\approx \lambda$); $\eta_{FS}(L)$ as free space propagation efficiency, corresponding to the beam divergence.

3.2.2 Gaussian Beam Propagation

Alternatively the Gaussian beam propagation model uses the beam radius or half spot size w_0 and $w(L)$ to describe the wave front. Due to the divergence, the beam radius increases with distance L from the transmitter. The free space transmission efficiency $\eta_{FS}(L)$ of the link budget will be basically replaced by $w(L)^2/w_0^2$ representing the beam divergence. The distribution of the irradiance $I(r, L)$ of the Gaussian beam with the beam radius $w(L)$ at a distance L from the transmitter, with r as lateral radius from the beam axis, L_R as Rayleigh length, and D_R as receiver aperture diameter, can then be written as

$$\frac{I(r, L)}{I(r, 0)} = \frac{w_0^2}{w(L)^2} \cdot \exp\left(-\frac{2 \cdot r^2}{w(L)^2}\right) \cdot \eta_T \cdot \eta_{Atm}(L) \cdot \eta_P \tag{3.2}$$

$$w(L) = w_0 \cdot \sqrt{1 + \left(\frac{L}{L_R}\right)^2} \tag{3.3}$$

$$L_R = \frac{\pi \cdot w_0^2}{\lambda} \tag{3.4}$$

$$P(r, L) = I(r, L) \cdot \pi \cdot D_R^2 \cdot \eta_R \tag{3.5}$$

3.2.3 Transmitter and Receiver

Coherent lasers for free space optical communications are either high power solid state lasers like Ti:Sapphire lasers pumped by an argon laser, or a classical commercial III/V semicondutor InGaAsP laser diode (approximately 10 mW). In combination with laser diodes, also EDFAs (Erbium Doped Fiber Amplifiers) can be used to achieve approximately 25 dB gain.

The silicon photodiode or the InGaAsP photodiode, also used in fiber optical communication networks, can be used for FSOC. For the extreme distance of 1–3 AU (astronomical unit, approximately 150,000,000–450,000,000 km) only few photons per pulse will arrive at the receiver. For this case an avalanche photodiode (APD) with built-in amplification in the Geiger mode can be used, and superconducting single photon counter, in order to increase the receiver responsivity significantly. However due to the deadtime of the APD in Geiger mode, the slot width of the PPM symbols cannot be chosen arbitrarily small, so that the total bitrate is also limited.

In Chap. 4 transmitters and receivers for different distances for uplink and downlink will be discussed more in detail.

3.2.4 Telescopes

At the transmitter, the laser beam will be collimated to the telescope aperture. Due to the Gaussian distribution of the fundamental mode TEM_{00}, the aperture is never illuminated 100% perfectly, leading to the above-mentioned transmitter efficiency of approximately $\eta_T \approx 70\%$. Depending on the size of the aperture, a corresponding gain can be achieved. Due to the limited space on a spacecraft or satellite, the aperture size will be limited. On the other hand, the ground station receiver telescope can be designed large enough to achieve the best received beam power and the best possible signal-to-noise ratio (SNR).

In general, the received noise is caused by the blue sky background, the Earth background, the sun light or the irradiance from other stars or planets.

3.2.5 Absorption and Scattering in the Atmosphere

During free space optical wave propagation through the atmosphere, some photons will be absorbed by molecules and electrons in the atmosphere. The absorption by molecules is in the infrared spectral range, whereas the absorption by electrons is mostly in the ultraviolet spectral range. Besides these absorptions water molecules show typical resonances at different wavelengths, leading to high absorption loss and almost 0% transmittance (see for example [8]). But fortunately the atmosphere shows a high transmittance of $70\% \le \tau(L) \le 90\%$ at several wavelengths, allowing laser optical transmission. Some typical wavelengths of 800 nm - 850 nm, 1064 nm, or 1550 nm can be used for deep space optical communications.

In free-space optical transmission absorption and scattering losses occur similarly to wired optical data transmission and these losses can vary in magnitude depending on the wavelength. The main causes are the molecules of water, carbon dioxide, and ozone.

Table 3.1 Molecular absorption depending on wavelength

Wavelength (nm)	Molecular absoprtion (dB/km)
550	0.13
690	0.01
850	0.41
1550	0.01

These losses can be described in terms of an attenuation coefficient which is the sum of the absorption and scattering coefficients of aerosols and molecular components of the atmosphere.

The optical signal is affected by molecules in the atmosphere. Therefore, it is reasonable to assume that different propagation conditions prevail under different weather conditions such as snow, rain, or fog. Fog, in particular, has the greatest influence, since the particle size roughly corresponds to that of the wavelengths of interest. The typical molecular absorption depending on wavelength is shown in Table 3.1.

To be able to describe the different weather conditions, atmospheric visibility was defined as a measure. However, this requires the relationship between visibility and attenuation. For this, the term "specific attenuation" is used [8, 9].

3.2.6 Pointing Errors, Spacecraft Jitter, and Point Ahead Errors

Other critical effects, like pointing errors, spacecraft jitter, and point ahead errors, are not discussed explicitly in this book, but can be implicitly considered by the pointing efficiency η_P and spacecraft transmitter efficiency η_T for link budget estimations.

3.2.7 Turbulence, Scintillation, and Beam Wander

Besides Rayleigh scattering and Mie scattering at the molecular structures in the atmosphere, there are also turbulences caused by the motion of the atmosphere and clouds, with dimensions of a few millimeters to eddies of some meters. All these turbulences cause a stochastic fluctuation of the refractive index of the atmosphere, therefore phase shifts and refraction. The fluctuation of the refractive index of the eddies with sizes l_0 of few millimeters is called small-scale scintillation, whereas the large-scale fluctuation L_0 of the refractive index leads to refraction of the beams in a stochastic manner, so that the beam changes the propagation direction back and forth. This effect is also called beam wander. Both small-scale and large-scale scintillation lead to a timely stochastic variation of the beam irradiance or the received power. This has to be taken into account.

The turbulences can be described by the Kolmogorov theory of turbulences (see for example [8]) and characterized by the so-called structure parameter of the refractive index fluctuation in [$m^{-2/3}$] which depends on the height above the sea level h in [m], the wind velocity v in [m/s], the temperature $T(\mathbf{R})$ in [K] and the pressure $P(\mathbf{R})$ in [mbar], where $\mathbf{R}$ is the vector pointing from the coordinate center to the position in the atmosphere under consideration. The normalized refractive index for optical waves can be described by the following equation with the units for v in [m/s] and h in [m] as well as λ in [μm][8] with the second term describing the fluctuation or scintillation effect

$$n(\mathbf{R}) = 1 + 77.6 \cdot 10^{-6} \cdot \left(1 + 7.52 \cdot 10^{-3} \cdot \lambda^{-2}\right) \cdot \frac{P(\mathbf{R})}{T(\mathbf{R})} \tag{3.6}$$

For $\lambda = 0.5\ \mu$m and $\lambda = 1.55\ \mu$m the refractive index fluctuation will be

$$n(\mathbf{R}) = 1 + 79.9 \cdot 10^{-6} \cdot \frac{P(\mathbf{R})}{T(\mathbf{R})} \tag{3.7}$$

$$n(\mathbf{R}) = 1 + 77.8 \cdot 10^{-6} \cdot \frac{P(\mathbf{R})}{T(\mathbf{R})} \tag{3.8}$$

Widely used wavelengths are 0.84 μm, 1.064 μm and 1.55 μm. In this book we assume one single formula (3.9) for the complete wavelength range for the optical laser communications $\lambda = 0.5\ \mu$m - 1.55 μm with sufficient accuracy, also given by [9], which is exact for 1.0 μm

$$n(\mathbf{R}) = 1 + 79 \cdot 10^{-6} \cdot \frac{P(\mathbf{R})}{T(\mathbf{R})} \tag{3.9}$$

All the parameters change permanently depending on wind velocity and height, and can be calculated by using an empirical formula, i.e. with $A = 1.7 \cdot 10^{-14}\ \mathrm{m}^{-2/3}$ for a typical wind velocity $v = 21\,\mathrm{km/h} \approx 5.83\,\mathrm{m/s}$ [8]

$$\begin{aligned} C_n^2(h, v) = {} & 0.00594 \cdot \left(\frac{v}{27}\right)^2 \cdot \left(10^{-5} h\right)^{10} \cdot \exp\left(\frac{-h}{1000}\right) \\ & + 2.7 \cdot 10^{-16} \cdot \exp\left(\frac{-h}{1500}\right) + A \cdot \exp\left(\frac{-h}{100}\right) \end{aligned} \tag{3.10}$$

The lower the wind velocity and the higher the altitude, the lower the value of C_n^2 will be. Especially for the case if the altitude of the ground station is chosen higher, rather than at the ground level, so that the effective distance between the spacecraft and the ground station becomes significantly shorter. The C_n^2 values can be significantly reduced also by considering the elevation angle.

There are many software programs for a Hufnagel-Valley (H-V) Model calculation or a graphical diagram [9] can be used to find the typical values $C_n^2 \approx 10^{-14} - 10^{-13}\mathrm{m}^{-2/3}$ at ground or sea level, also called strong turbulence; $C_n^2 \approx 10^{-16}\mathrm{m}^{-2/3}$ at a height of about 1 km above sea level; $C_n^2 \approx 10^{-17}\mathrm{m}^{-2/3}$ at $h = 3\,\mathrm{km}$ above sea level, with a wind velocity

v = 21 km/h, a rather weak turbulence; $C_n^2 < 10^{-18} - 10^{-17}\mathrm{m}^{-2/3}$ at h = 3 - 10 km above sea level, with weak or no wind; $C_n^2 < 10^{-19} - 10^{-18}\mathrm{m}^{-2/3}$ at $h >$ 10 km above sea level. The near Earth troposphere changes gradually from approximately 6 - 15 km to tropopause, depending on the latitude.

The Rytov variance for spherical waves can be used to estimate the standard deviation of the irradiance or the field amplitude with units for k in [m^{-1}] and L in [m]

$$\sigma_I^2(L) = 0.5 \cdot C_n^2 \cdot k^{7/6} \cdot L^{11/6} \tag{3.11}$$

3.3 Conclusions

The simulation methodology for free space optical communications (FSOC) and deep space optical communications (DSOC) is presented in this book. By using the above-mentioned link budget algorithm and the Rytov variance, the link scenarios (a) between the near Earth GEO satellite ARTEMIS and (b) between the Lagrange point L5 and the optical ground station OGS by considering the atmospheric scintillations, can be investigated and simulated in Chaps. 5 and 6.

Since the structure function for the refractive index is changing over the large distance, also varying due to changes of temperature, pressure, wind, azimuth, and elevation angles, one cannot find an analytical set of parameters which is valid for all scenarios. The comparison shows that the simulation models for the link budget and the scintillation estimation are feasible for the investigations of free space and deep space optical communications. This investigation helps to find a reliable empirical data set for each project with typical constellations, daytimes, and typical weather conditions. In the near future we will be working on link performance analyses for downlink and uplink between Earth ground station and lunar orbits, and between Earth ground station and other planets or asteroids by using realistic design parameters.

References

1. H. Hemmati, A. Biswas, I. B. Djordjevic, "Deep Space Optical Communications: Future Perspectives and Applications", Proceedings of IEEE, vol. 99, No. 11, November 2011.
2. Marc Sans and Zoran Sodnik, Igor Zayer and Robert Daddato, "Design of the ESA Optical Ground Station for Participation in LLCD", Proc. International Conference on Space Optical Systems and Applications (ICSOS) 2012, 3-1, Ajaccio, Corsica, France, October 9-12, 2012.
3. Zoran Sodnik and Reinhard Czichy, "Design Data Summary of the ESA Optical Ground Station for In-Orbit Check-Out of Laser Communication Payloads and for the Observation and Registration of Space Debris". ESA Specifications for the Artemis OGS OPALE project. Ref.: TEC-MMO/267/ZS. May 2, 2005.
4. Internal ESA Artemis OGS OPALE Measurement Results. 2003-2004.

5. J. Mendenhall, L. Candell, P. Hopman, G. Zogbi, D. Boroson, D. Caplan, C. Digenis, D. Hearn, and R. Shoup, “Design of an Optical Photon Counting Array Receiver System for Deep-Space Communications”, Proceedings of IEEE, vol. 95, Nr. 10, October 2007.
6. Zoran Sodnik, Hans Smit, Marc Sans, Dirk Giggenbach, Peter Becker, Ramon Mata-Calvo, Christian Fuchs, Igor Zayer, Marco Lanucara, Klaus-Jürgen Schulz, Johannes Widmer, Felix Arnold, Martin Mosberger, Angel Alonso, Iciar Montilla, “Results from a Lunar Laser Communication Experiment between NASA’s LADEE Satellite and ESA’s Optical Ground Station”, Proc. International Conference on Space Optical Systems and Applications (ICSOS) 2014, S2-1, Kobe, Japan, May 7-9, 2014.
7. Zoran Sodnik, Clemens Heese, Pantelis-Daniel Arapoglou, Klaus-Júrgen Schulz, Igor Zayer, Robert Daddato, Stefan Kraft, “Deep-Space Optical Communication System (DOCS) for ESA’s Space Weather Mission to Lagrange Orbit L5”. IEEE International Conference on Space Optical Systems and Applications (ICSOS), November 14-16, 2017.
8. Larry C. Andrews, Ronald L. Phillips, Cynthia Y. Hopen, “Laser Beam Propagation through Random Media”, SPIE Press, Bellingham, Washington USA, 2005.
9. Larry C. Andrews, Ronald L. Phillips, Cynthia Y. Hopen, “Laser Beam Scintillation with Applications”, SPIE Press, Bellingham, Washington USA, 2001.
10. S.-P. Chen: Investigations of free space and deep space optical communication scenarios. (2021) CEAS Space Journal, 14, 357–364. https://doi.org/10.1007/s12567-021-00388-y
11. Tobias Siegel, Shun-Ping Chen: Investigations of Free Space Optical Communications unter Real-World Atmospheric Conditions. Wireless Personal Communications, Springer Nature. https://doi.org/10.1007/s11277-020-07724-1, August 12, 2020.
12. E. D. Aboagye, S.-P. Chen: Deep space optical communications (DSOC) downlink simulation with varying PPM order. Optical and Quantum Electronics, 53, 583 (2021). https://doi.org/10.1007/s11082-021-03232-z
13. S.-P. Chen: Fundamentals of Information and Communication Technologies. Cambridge Scholars Publishing (2020).
14. F. Schüßler, T. Stieglitz, P. Thielke, S.-P. Chen (Supervisor): FSOC (Free Space Optical Communications) und DSOC (Deep Space Optical Communications). IWP Team Research Project (German). Faculty of Electrical Engineering and Information Technology, h-da Darmstadt University of Applied Sciences. June 19, 2024.
15. W. Marshall und B. Burk: Received Optical Power Calculations for Optical Communications Link Performance Analysis, NASA, Technical Report, 1986.
16. R. Long: Atmospheric attenuation of ruby lasers, IEEE, 1963. https://doi.org/10.1109/PROC.1963.2294.
17. A. S. Jursa: Handbook of Geophysics and the Space Environment. Air Force Geophysics Laboratory Air Force Systems Command of the United States Air Force. 1985.
18. M. Rouissat, R. . Borsali und M. Chick-Bled: Free Space Optical Channel Characterization and Modeling with Focus on Algeria Weather Conditions. International Journal of Computer Network and Information Security, Vol. 4, Pages. 17–23, 2012.
19. M. C. Alnaboulsi und F. de Fornel: Fog attenuation prediction for optical and infrared waves. Optical Engineering, 2004. https://doi.org/10.1117/1.1637611.
20. I. I. Kim, B. McArthur und E. J. Korevaar: Comparison of laser beam propagation at 785 nm and 1550 nm in fog and haze for optical wireless communications. SPIE Journal, 2001.

4 Transmitter and Receiver for Free Space and Deep Space Optical Communications

4.1 Introduction

This book is related to research projects [1–4, 6–8] dealing with Free Space Optical Communication (FSOC) and Deep Space Optical Communication (DSOC). The research projects were performed at the Institute of Communication Technologies of the Department of Electrical Engineering and Information Technologies at Darmstadt University of Applied Sciences. These projects investigated the design and implementation of photon-based communication systems for both uplink and downlink channels in space, based on the publications of well-known research institutions and teams [9–36].

It focuses on the representative scenarios: FSOC for terrestrial communications or near-Earth satellites links, inter-satellite optical links, and DSOC for lunar, Earth-to-Lagrange points, and interplanetary missions.

Free Space Optical Communication (FSOC) systems are characterized by free space as the communication channel between transceivers, also called line-of-sight (LOS) links for successful transmission of optical signals. The channel can be atmosphere, space, or vacuum whose characteristics determine the transmission. These characteristics have to be taken into account for designing dependable and effective communication systems. Using FSOC technology, data is transmitted by propagation of light through atmospheric or space communication channels, allowing optical connectivity. FSOC offers a high data rate and meets the tremendous adding demand for broadband data transmission substantially driven by Internet access and HDTV broadcasting services.

For several decades, deep space communication has relied on conventional radio frequency (RF) bands, including S-band (2–4 GHz), X-band (8–12 GHz), and Ka-band (26–40 GHz). These frequency ranges have been instrumental in facilitating reliable communication with spacecrafts across the solar system at NASA's Deep Space Network (DSN) and the European Space Agency's ESTRACK stations.

S.-P. Chen, *Fundamentals of Free Space and Deep Space Optical Communications*,
https://doi.org/10.1007/978-3-032-35826-4_4

However, RF communication faces several limitations. Firstly, the available bandwidth is relatively limited, which restricts the achievable data rates, especially as spacecrafts generate more high-resolution images and scientific data. Secondly, RF systems necessitate the implementation of substantial antennas on both the spacecrafts and the ground stations to counteract the broad beam divergence characteristic of lower frequencies. As a result, mass, power consumption, and cost of the mission are increased. The RF spectrum is subject to stringent regulations and experiences significant congestion with numerous users and services competing for limited frequency assignments. These challenges underscore the necessity for a transition toward more efficient communication technologies, such as optical links.

The rapid growth of data demands in space missions has highlighted the limitations of conventional RF communication systems, particularly in terms of bandwidth, antenna size or antenna aperture, and spectrum congestion. As a response, optical communication technologies have emerged as a promising alternative, offering higher data throughput, reduced beam divergence, reduced weight, and efficient spectrum utilization.

The book reviews key enabling technologies, including semiconductors, fiber, and solid-state laser transmitters, as well as advanced receiver architectures employing avalanche photodiodes, superconducting nanowire detectors, and photon counting techniques. In addition, it outlines the principles of system architecture, channel impairments, and link budgeting, providing a structured framework for evaluating performance and feasibility.

By consolidating technical trade-offs, system requirements, and emerging solutions, this chapter offers a comprehensive overview of how optical communication can meet the evolving challenges of space exploration and lay the foundation for future high-capacity, interplanetary networks.

This chapter also provides a concise and rigorous assessment of contemporary transmitter and receiver technologies for optical free space communications, and it covers both space-to-ground downlink and ground-to-space uplink connections using three operational contexts as reference scenarios: communications with satellites in Earth orbit, deep space downlinks from interplanetary probes, and operations near the Earth-Sun Lagrange points. It documents the hardware classes most often used in practice and describes the subsystem functions that support transmission and reception under realistic conditions such as ambient background light, atmospheric effects, and the demands of precise pointing and tracking.

The review summarizes established laser architectures and amplifier solutions for transmitters and outlines the mission-driven trade-offs that govern their selection and integration into system designs. It also compares available detector technologies for receivers and discusses system-level implications, including thermal management interface complexity and sensitivity performance, so that equipment choices can be weighed against operational constraints.

The Lunar Laser Communication Demonstration (LLCD) project, conducted by NASA, in conjunction with the ARTEMIS optical ground station experiments, which were carried out by the European Space Agency (ESA), has demonstrated the viability of laser links for utilization in both near-Earth and lunar missions. Successes in the field have led to the development of ambitious missions such as NASA's Psyche and ESA's DOCS at Lagrange L5, which aim to extend optical communication to distances in deep space measured in astronomical units (AU).

Optical links are increasingly replacing RF for deep-space missions, thanks to their narrow beam divergence and high data rates with compact terminals. ESA's ARTEMIS/OGS campaign validated FSOC in Earth orbit, while NASA's Psyche mission demonstrated DSOC across interplanetary distances, achieving 267 Mbps over a distance of 19 million miles and reliable links beyond 2 AU. These results confirm the feasibility of DSOC systems and provide models for future missions.

Free-Space Optical Communication (FSOC) transmits digitally modulated laser light through air or vacuum instead of glass fiber. A narrow optical beam typically at 850, 1064, or 1550 nm is generated by a semiconductor, fiber, or DPSS laser, modulated (e.g., OOK, power-efficient PPM, or higher- order coherent schemes like QPSK/QAM), and transmitted by a telescope/beam director toward a distant terminal. The receiver uses a collecting telescope, tight optical filtering, and sensitive photodetectors (Si or InGaAs APDs; for extreme links, superconducting nanowire detector) followed by low-noise Transimpedance Amplifier (TIA), clock/data recovery, and strong Forward-Error-Correction (FEC) LDPC/Turbo to reconstruct the bitstream. Because optical wavelengths are about 1 μm, beams have micro-radian divergence and very high "antenna" gain (the "antenna" apertures are lenses or mirrors much larger than the wavelength), enabling multi-Gbps data rates with compact, low-mass terminals and operation in license-free spectrum. These traits make FSOC well-suited for satellite downlinks, inter-satellite crosslinks, aircraft-to-ground links, and deep-space missions.

As illustrated in Fig. 2.1 (Free Space Optical Link Model), the process begins with the laser transmitter (Laser TX) generating an optical signal. This signal is then expanded and collimated by the transmitting telescope (TX Telescope Gain) to ensure efficient propagation over long distances. The beam travels through the FSOC channel, which represents the free space medium and is subject to impairments such as atmospheric absorption, scattering, turbulence, and pointing errors. At the receiver side, the receiving telescope (RX Telescope Gain) collects the incoming optical beam and focuses it onto the detector. Finally, the photodiode (APD or PIN detector) converts the optical photons into electrical signals, which are processed and decoded to recover the transmitted information. This chain illustrates the fundamental principle of FSOC links, where high-power lasers, precise optics, and sensitive detectors work together to enable high-data-rate communication through free space.

Depending on the transmission path, i.e. either from the Earth ground station upward to the spacecraft or downward from the spacecraft to the Earth ground station, the transmission path will be called either Uplink or Downlink.

4.2 FSOC and DSOC Uplink Transmitter

FSOC and DSOC Uplink Transmitter (see also Table 4.1) involves transmitting a laser beam from a ground station towards a satellite or an airborne platform. The choice of transmitter and receiver technologies plays a critical role in determining the overall link performance, especially given the challenges of atmospheric turbulence, pointing accuracy, and low photon budgets in space communications.

Table 4.1 contains the technology of the FSOC and DSOC Uplink Transmitter described in the following sections, additionally the mission examples, power, wavelength, TX-aperture, and some important remarks.

The transmitter in an optical uplink system must generate a highly directional, stable, and power-efficient optical signal to bridge the extremely long distance, enabling the successful reception of the signal, in combination with the extremely sensitive receiver. Several types of laser sources and modulation formats are used for the Uplink transmitter:

4.2.1 High-power Semiconductor Lasers

High-power semiconductor lasers are widely used in optical uplink systems due to their compact size, electrical efficiency, and commercial availability. They typically provide output powers in the range of 1–10 W and operate at wavelengths such as 1064 and 1550 nm. Among these, the 1550 nm wavelength is particularly attractive for space communication because it is compatible with existing telecom components and is classified as eye-safe for ground operations, reducing risks during terrestrial use.

Despite these advantages, semiconductor lasers generally suffer from lower beam quality compared to fiber or diode-pumped solid-state (DPSS) lasers. This limitation often necessitates the use of beam-shaping optics to ensure the optical beam maintaining the required quality and directionality over long-distance free-space links.

4.2.2 Fiber Lasers

Fiber lasers are another important transmitter technology for optical uplinks, offering output powers that can exceed 100 W while maintaining excellent, near diffraction-limited beam quality. They typically operate within the 1030–1070 nm wavelength range and are recognized for their high optical efficiency, robustness, and thermal stability. Fiber lasers are also highly scalable, which makes them attractive for applications that require extremely high power levels. Their superior beam quality ensures minimal divergence, a key requirement for reliable ground-to-satellite links. However, a drawback lies in their operating wavelengths, which may overlap with atmospheric absorption bands. This makes careful wavelength selection essential in order to minimize transmission losses and ensure optimal link performance.

Due to the high power and superior beam quality to overcome free-space losses, fiber lasers are well suited for ground uplink transmitters in FSOC applications.

4.2.3 Diode-Pumped Solid-State (DPSS) Lasers

DPSS lasers, such as Nd:YAG lasers operating at 1064 nm, are well established in space communication applications due to their high coherence, narrow linewidth, and excellent long-term stability. These properties make them particularly suitable for coherent communication systems where precise spectral characteristics are required. DPSS lasers offer a good balance between power output and optical quality, providing stable and reliable operation under demanding conditions.

4.2.4 Quantum Cascade Lasers (QCLs)

Quantum cascade lasers operate in the mid-infrared range, typically between 3–12 μm, and provide unique advantages for uplink communication. By working in this spectral band, QCLs can reduce the effects of atmospheric scattering and absorption compared to shorter wavelengths, enabling more robust performance under challenging environmental conditions. They are particularly promising for niche applications such as secure communication links or specialized uplinks in turbulent atmospheric channels. However, current limitations include relatively low wall-plug efficiency and the need for active cooling, which restricts their widespread use in large-scale space communication systems.

As research progresses, QCLs are expected to become a more practical solution for certain high-performance uplink applications.

4.2.5 Overview of Uplink Transmitter Features

- High-power semiconductor lasers: compact, efficient, telecom/eye-safe 1550 nm options; good size, weight and power (SWaP) features for especially for flight terminals;
- Fiber lasers: 100 W+, near-diffraction-limited beams and excellent stability, perfect for long-range, tight-divergence uplinks;
- DPSS (Nd:YAG 1064 nm): narrow linewidth, high coherence, and proven reliability, ideal for coherent links and stable operation;
- QCLs (3–12 μm): mid-IR cuts scattering/absorption for harsh atmospheres, niche today due to cooling/efficiency needs;

- Mid-IR Wavelengths (3–12 μm): Operate in atmospheric transmission windows with reduced scattering and absorption;
- High Power and Tunability: QCLs provide broad wavelength coverage, suitable for specialized FSOC links;
- Compact and Reliable: Solid-state semiconductor design with no moving parts, ideal for space environments;
- Applications: Useful in conditions where standard telecom wavelengths (1064, 1550 nm) face higher atmospheric losses (e.g. dusty or humid environments).

4.3 FSOC and DSOC Uplink Receiver

FSOC and DSOC Uplink Receiver (see also Table 4.2) plays a crucial role in optical uplink systems, as it must detect extremely weak optical signals after they have passed through the atmosphere and experienced attenuation, scattering, and turbulence. The choice of the photodetector directly affects the sensitivity, noise performance, and overall reliability of the communication link. Several types of detectors are commonly employed, each with unique advantages and limitations.

Table 4.2 contains the technology of the FSOC and DSOC Uplink Receiver described in the following sections, additionally the mission examples, photon detection, wavelength, RX-aperture and some important remarks.

4.3.1 Silicon Avalanche Photodiodes (Si-APDs)

Si-APDs are widely used in optical receivers operating in the visible to near-infrared spectrum, covering wavelengths between 400 and 1000 nm. They offer high internal gain and low noise characteristics, making them effective for detecting weak signals in this range. Their mature technology and relatively low cost add to their attractiveness. However, their spectral response is limited, which makes them unsuitable for telecom wavelengths above 1000 nm.

Features:

- Useful for short-range FSOC links (LEO satellites, UAVs) where wavelengths less than 1000 nm are practical;
- Provide a low-cost, reliable solution for initial FSOC deployments or experimental setups;
- Enable efficient photon detection in visible/NIR bands, supporting applications where eye-safety or atmospheric transparency is favorable;

4.3.2 Indium Gallium Arsenide Avalanche Photodiodes (InGaAs APDs)

InGaAs APDs extend detection capabilities into the 1000–1700 nm range, covering the key wavelengths of 1064 nm and 1550 nm that are widely used in optical communication systems. They provide excellent sensitivity and are optimized for the telecom bands, where atmospheric transmission windows are most favorable. Due to their performance in these spectral regions, InGaAs APDs are considered the workhorse of ground-to-satellite and deep-space optical communication systems.

Features:

- Ideal for 1550 nm FSOC links, which are eye-safe and experience low atmospheric loss;
- Provide enhanced detection performance for medium-to-long-range optical links (LEO, GEO, interplanetary);
- Support higher data rates with lower background noise compared to visible-band detectors;
- Widely used in current telecom infrastructure, thus their technology is mature and easily space qualifiable.

4.3.3 Photomultiplier Tubes (PMTs)

PMTs remain indispensable in scenarios requiring ultra-high sensitivity. Capable of detecting single photons, PMTs are well suited for photon-starved environments, such as deep-space optical links. Their ability to provide photon-counting functionality makes them invaluable in scientific and exploratory missions. However, they are bulky, fragile, and require high-voltage operation, which can make integration into compact, spaceborne systems challenging.

Features:

- Useful in photon-starved FSOC link (long distances, weak signals);
- Enable single-photon detection for experimental or specialized short-wavelength optical communication systems;
- Valuable in proof-of-concept deep-space optical tests, though less common in modern telecom-band FSOC;
- Serve as benchmark detectors before transitioning to more compact APDs or Superconducting Nanowire Single-Photon Detectors (SNSPDs).

4.3.4 Silicon Photomultipliers (SiPMs)

SiPMs are emerging as a modern alternative to PMTs, offering single-photon sensitivity in a compact and solid-state form factor. They operate at lower voltages, are mechanically robust, and are more easily integrated into modern receiver systems. SiPMs are particularly attractive for photon-counting applications where size, weight, and power efficiency are critical considerations. Although still under active research for space-qualified communication systems, they hold significant promise as the next-generation detector technology for optical uplinks.

Features:

- Provide single-photon sensitivity in a small, lightweight package—ideal for space applications;
- Robust and resistant to shock/vibration, unlike fragile PMTs;
- Useful for short-range or atmospheric FSOC links (e.g., aircraft-to-satellite, LEO downlinks);
- Low cost and scalability make them attractive for multi-detector receiver arrays;
- Although limited in telecom bands (wavelength larger than 1000 nm), they are a promising option where visible/ Near Infrared (NIR) FSOC is applied.

4.4 FSOC and DSOC Downlink Transmitter

FSOC and DSOC Downlink Transmitter (Table 4.3) is located onboard a satellite or airborne platform and must be carefully engineered to withstand the harsh conditions of space while maintaining reliable optical performance. Space-qualified laser sources are designed with radiation hardening, thermal stability, and long operational lifetimes to ensure robust communication between spacecraft and ground terminals.

Table 4.3 contains the technology of the FSOC and DSOC Downlink Transmitter described in the following sections, additionally the mission examples, power, wavelength, TX-aperture and some important remarks.

4.4.1 Space-Grade Semiconductor Lasers

Space-Grade Semiconductor Lasers are among the most widely used sources due to their compact size, high efficiency, and straightforward integration with spacecraft systems.

They are radiation-hardened to endure the space environment and typically provide output powers in the range of 1–5 W. These lasers are lightweight and reliable, making them suitable for small satellites and missions where size, weight, and power (SWaP) constraints are critical.

These semiconductor lasers can also be considered for DSOC as they are compact, radiation-hardened, and efficient and suitable for spacecraft with limited power and harsh radiation environments.

4.4.2 Master Oscillator Power Amplifier (MOPA)

The MOPA combines a narrow-linewidth master oscillator with a high-power optical amplifier. This configuration enables high output power while preserving spectral purity and coherence, which are essential for coherent communication links and advanced modulation formats. The MOPA approach provides both flexibility and scalability, making it a preferred choice for high-data-rate satellite downlinks.

4.4.3 Quantum Dot Lasers (QDL)

QDLs are an emerging technology offering improved temperature stability compared to conventional semiconductor lasers. Operating in the 1300–1550 nm wavelength range, they combine the benefits of low threshold currents, high efficiency, and resilience to temperature fluctuations, which is highly advantageous in the variable thermal conditions of space.

QDLs can be considered for DSOC as they have excellent temperature stability, efficiency, reduced thermal control needs and enhanced reliability in space.

4.4.4 Distributed Feedback (DFB) Lasers

DFB lasers are also widely used in downlink transmitters, providing single-mode operation with narrow linewidths. Their spectral stability and low noise performance make them well suited for coherent communication schemes and dense-wavelength-division-multiplexing (DWDM) applications. In addition, their mature technology and proven reliability in terrestrial telecom networks have facilitated their adaptation for space communications, ensuring stable operation over extended mission durations.

DFB lasers are considered in DSOC because of their single-mode, narrow linewidth output which enables precise, coherent transmission needed for high data rate deep-space links.

4.5 FSOC and DSOC Downlink Receiver

In optical downlink communication, the ground terminal receiver or FSOC and DSOC Downlink Receiver (see also Table 4.4) is tasked with detecting extremely weak optical signals transmitted from satellites or airborne platforms. Atmospheric turbulence, beam spreading, and pointing uncertainties make receiver design especially critical for ensuring reliable link performance. To address these challenges, several types of advanced photodetectors are used in ground terminals, each tailored to balance sensitivity, noise performance, and spatial coverage.

Table 4.4 contains the technology of the FSOC and DSOC Downlink Receiver described in the following sections, additionally the mission examples, wavelength, RX-aperture and some important remarks.

4.5.1 Large-Area Avalanche Photodiodes (APDs)

Large-Area Avalanche Photodiodes are frequently employed in downlink systems because their increased active area allows for the collection of more photons, thereby reducing the impact of pointing errors and beam wander. While larger active regions can introduce additional capacitance and noise, the trade-off is beneficial in scenarios where maintaining stable optical alignment is difficult. These detectors provide high sensitivity and robustness, making them well suited for satellite-to-ground links.

Features:

- Eases alignment challenges: Larger active area means the optical beam does not need to be focused precisely, critical in ground-to-satellite and long-range links where beam wander occurs;
- Higher collection efficiency: Captures more incoming photons, improving link margin;
- Reduced pointing errors: Makes the system more robust under turbulence and spacecraft vibration;
- Compact and space-qualified: Solid-state solution with heritage in telecom and satellite systems.

4.5.2 Multi-element Detector Arrays

Multi-Element Detector Arrays offer another approach by providing spatial diversity in photon detection. By distributing the incoming optical beam across multiple detector elements, they can mitigate the effects of atmospheric scintillation and beam distortions, thereby improving link availability and stability. Detector arrays also enable advanced signal processing techniques such as spatial combining and adaptive beam tracking, further enhancing receiver performance under variable atmospheric conditions.

Features:

- Improved pointing tolerance: Reduced need for ultra-precise beam alignment, since even if the spot shifts, a part of the array still captures the signal;
- Scintillation mitigation: Atmospheric turbulence causes beam wander and intensity fluctuations. Arrays smooth out these effects;
- Spatial diversity: Multiple detectors increase link reliability and reduce fades;
- Scalable: Can be expanded for higher sensitivity and larger collection areas.

4.5.3 Superconducting Nanowire Single-Photon Detectors (SNSPDs)

SNSPDs represent the state of the art in photon detection for optical communications. Operating at cryogenic temperatures, they provide near-unity quantum efficiency, extremely low dark counts, and unparalleled timing resolution. These features make SNSPDs the ultimate choice for ultra-sensitive downlink applications, such as deep space optical communications, where the received photon flux is extremely low. However, their reliance on cryogenic cooling systems increases complexity and limits their current widespread deployment.

Features:

- Ultimate sensitivity: Detects individual photons, crucial for deep-space downlinks where signals are extremely faint;
- High quantum efficiency: Maximizes photon information efficiency, supporting higher data rates;
- Ultra-low noise: Reduces error rates, enabling long-distance reliable links;
- Fast response: Supports high-speed modulation schemes like PPM at photon starved conditions;
- Future-ready: Compatible with quantum key distribution (QKD) in space, enabling secure interplanetary networks.

4.5.4 Hybrid Photomultiplier Detectors (HPDs)

HPDs combine the advantages of conventional photomultipliers with solid-state technology to achieve low noise levels and high quantum efficiency. They are particularly effective in environments where both sensitivity and noise suppression are critical for link performance. HPDs offer a compromise between the extreme sensitivity of SNSPDs and the practicality of APDs, making them a versatile choice for certain ground terminal applications.

4.5.5 Overview of the Downlink Receiver Features

Si APDs (400–1000 nm):

- Low noise;
- fast timing;
- simple linear receivers;
- great for short/medium links at 850–905 nm.

InGaAs APDs (1000–1700 nm):

- High quantum efficiency at 1064/1550 nm;
- the telecom bands;
- so they are the default for atmospheric FSOC (PPM + narrow filters).

PMTs:

- Ultra-high gain photon counting at visible wavelengths;
- useful for short wavelength demos but mismatched to 1550 nm and bulky for space.

SiPMs (less than 900 nm):

- Compact solid-state photon counters with excellent timing;
- good for airborne/terrestrial links and acquisition sensors; not suitable 1550 nm;
- High sensitivity: Suitable for detecting very weak optical signals, critical in long-distance space links;
- Low noise operation: Improves signal-to-noise ratio (SNR) in photon-starved channels;
- Wide dynamic range: Can handle varying signal strengths without saturation;
- Better efficiency than PMTs: Higher quantum efficiency makes them more photon-economical;
- Reliable for ground stations: Used where cooling and stable infrastructure are available.

4.5.6 Comparison of Some Transmitters and Receivers for Uplink

Table 4.1 Uplink transmitters

Technology	Mission	Power	Wavelength (nm)	TX-Aperture (cm)	Remarks
Semiconductor Laser	ESA Artemis	15–300 mW	815–850	10–20	Compact, efficient, but lower beam quality
Er/Yb-doped Fiber Laser	ESA DOCS	4 × 600 W	1064	4 × 10	Telecom standard, excellent beam quality
EDFA MOPA	NASA/ESA LLCD	4 × 10/40 W	1550	4 x 10	Reliable, widely used in Lunar/GEO links

Table 4.2 Uplink receivers

Technology	Mission	Photon detection	Wavelength (nm)	RX-aperture (m)	Remarks
Si-APDs	Artemis	100–1000 photons/bit	800–900	0.15	High gain photon counting
InGaAs APDs	ESA DOCS	Few photons/slot	1064–1550	0.25	Telecom standard, excellent beam quality

4.5.7 Comparison of Some Transmitters and Receivers for Downlink

Table 4.3 Downlink transmitters

Technology	Mission	Power (W)	Wavelength (nm)	TX-Aperture	Remarks
Semiconductor Laser	ESA SILEX Artemis	1–5 W	850/1550 nm	12–15 cm	Compact, efficient for small satellite
EDFA MOPA	NASA/ESA LLCD	0.5	1550	10 cm	Compact, efficient for small satellite
MOPA	TESAT DLR LCT	1–2	1064	13.5 cm	High power, narrow linewidth, supports coherent links
DFB Lasers	TESAT LCT	1–5	1064	13.5 cm	Narrow linewidth, stable high bitrate, coherent
Quantum Dot Lasers	NICT, EU project	1	1300–1550	Chip-scale	Temperature stability, experimental stage

Table 4.4 Downlink receivers

Technology	Mission examples	Wavelength	RX-Aperture	Remarks
InGaAs APDs	ESA OGS	1064/1550 nm	1–1.55 m	Telecom bands
Large-area APDs	LLCD	1064 nm	40 cm	Reduced pointing error, higher noise trade-off
Detector Arrays	ESA SILEX	1064/1550 nm	1 m	Spatial diversity, mitigates scintillation and turbulence
SNSPDs	NASA DSOC LLCD	1550 nm	5 m	Ultra sensitive, cryogenic cooling required
HPD	JAXA in research	Visible - NIR	1 m	Balance of PMT sensitivity and solid-state robustness

References

1. S. Guezguez, E. D. Aboagye, K. Sauer, B. V. S. Sharma, P. Sukesh, S.-P. Chen (Supervisor): Deep Space Optical Communications (Link Performance by considering different models of attenuation and scintillation). Team Project. Institute of Communication Technologies, Darmstadt University of Applied Sciences. August 5, 2019
2. S. Srivastav, M. Adil, S. Shakir, R. Nasit, S. P. Meesala, S.-P. Chen (Supervisor): Analysis of Free Space Optical Communications (Link Performance by considering different models of attenuation and scintillation). Team Project. Institute of Communication Technologies, Darmstadt University of Applied Sciences. September 19, 2023
3. N. Manjunath, V. Dharamkari, M. Bharadwaj, A. A. Maurer, O. S. I. Shoeib, S.-P. Chen (Supervisor): DSOC Transmitter and Receiver for Uplink and Downlink. Team Project. Institute of Communication Technologies, Darmstadt University of Applied Sciences. August 26, 2025
4. R. G. R. Charles, S. Sukumaran, V. Jose, V. Sadashivpet, S.-P. Chen (Supervisor): Absorption and Scintillation of Free Space Optical Communications. Team Project. Institute of Communication Technologies, Darmstadt University of Applied Sciences. September 16, 2025
5. F. Schüßler, T. Stieglitz, P. Thielke, S.-P. Chen (Supervisor): Absorption and Scintillation of Free Space Optical Communications. IWP Team Research Project (in German). Faculty of Electrical Engineering and Information Technology, h-da Darmstadt University of Applied Sciences. June 19, 2024.
6. Tobias Siegel, Shun-Ping Chen: Investigations of Free Space Optical Communications unter Real-World Atmospheric Conditions. Wireless Personal Communications, Springer Nature. https://doi.org/10.1007/s11277-020-07724-1, August 12, 2020.
7. Aboagye, E. D., Chen, S.-P.: Deep space optical communications (DSOC) downlink simulation with varying PPM order. Optical and Quantum Electronics, 53, 583. https://doi.org/10.1007/s11082-021-03232-z (2021).
8. Chen, S.-P.: Investigations of free space and deep space optical communication scenarios. CEAS Space Journal, 14, 357–364. https://doi.org/10.1007/s12567-021-00388-y (2021).

9. Wollman, E. E., et al. (2024). An SNSPD-based detector system for NASA's Deep Space Optical Communications project. Optics Express, 32(27), 48185. https://opg.optica.org/oe/abstract.cfm?uri=oe-32-27-48185
10. Wollman, E. E., et al. (2024). An SNSPD-based detector system for NASA's DSOC project. arXiv:2409.02356. https://arxiv.org/abs/2409.02356
11. Wollman, E. E., et al. (2024). MIR-extended SNSPDs on SOI with near-unity efficiency to 3.5 μm. arXiv:2408.15959. https://arxiv.org/pdf/2408.15959
12. SPIE (2024). Superconducting nanowire single-photon detectors for laser communications (array, $<$ 50 ps jitter). Proc. SPIE 12877. https://ui.adsabs.harvard.edu/abs/2024SPIE12877E..0SG/abstract
13. NASA NTRS (2024). Testing of a Photon-Counting Optical Ground Receiver With SNSPDs and FPGA readout. https://ntrs.nasa.gov/citations/20240000366
14. SPIE (2024). DSOC technology demonstration pre-launch validation and performance tests with the Laser Test Evaluation Station (LTES). Proc. SPIE 12877. https://ui.adsabs.harvard.edu/abs/2024SPIE12877E..07V/abstract
15. Jarzyna, M., et al. (2024). Photon information efficiency limits in deep-space optical communications. Optical Engineering, 63(4), 041209. https://doi.org/10.1117/1.OE.63.4.041209
16. Optica CLEO (2024). 14 Bits per Photon Detection in the Photon-Starved Regime for Deep-Space Optical Communication.
17. NASA. Deep Space Optical Communications (DSOC)—Mission page with milestone timeline. https://www.nasa.gov/mission/deep-space-optical-communications-dsoc/
18. NASA/JPL (2024). NASA's Tech Demo Streams First Video From Deep Space via Laser. https://www.jpl.nasa.gov/news/nasas-tech-demo-streams-first-video-from-deep-space-via-laser
19. NASA/JPL (2023). DSOC Factsheet. https://www.jpl.nasa.gov/presskits/dsoc
20. NASA (2024). DSOC Fact Sheet (program goals and phases). https://www.nasa.gov/wp-content/uploads/2024/04/dsoc-factsheet.pdf
21. EdgeOptic (2025). DSOC first phase completed: 11 Tbit downlink from 460 million km, phase-2 plan. https://edgeoptic.com/deep-space-optical-communications-dsoc-mission-and-its-impact-on-our-future/
22. Avantier Inc. (2025). DSOC technology demonstration: 266 Mbps downlink and ground systems overview. https://avantierinc.com/resources/knowledge-center/deep-space-optical- communications/
23. Meenehan, S., et al. (2025). Operational results from the DSOC ground system. Proc. SPIE 13355, Free-Space Laser Communication XXXVII, 133550K. https://doi.org/10.1117/12.3041400
24. Marsili, F., et al. (2013). Detecting single infrared photons with 93
25. PhotonSpot (2024). SNSPD product datasheet. https://www.photonspot.com
26. Scontel (2024). SNSPD systems datasheet. https://www.scontel.ru
27. Excelitas (2024). SLiK APD Photodiode datasheet. https://www.excelitas.com
28. NKT Photonics (2024). Koheras BASIK and BOOSTIK fiber lasers. https://www.nktphotonics.com
29. opg.optica Optics Express — "An SNSPD-based detector system for NASA's Deep Space Optical Communications project" (2024): https://opg.optica.org/oe/abstract.cfm?uri=oe-32-27-48185
30. arxiv arXiv — Wollman et al., "An SNSPD-based detector system for NASA's DSOC project" (2024): https://arxiv.org/html/2409.02356v1
31. arxiv arXiv — MIR-extended SNSPDs on SOI with near-unity efficiency to 3.5 μm (2024): https://arxiv.org/pdf/2408.15959
32. ui.adsabs.harvard SPIE (Proc.) — "Superconducting nanowire single-photon detectors for laser communications" (array, $<$50 ps jitter) (2024): https://ui.adsabs.harvard.edu/abs/2024SPIE12877E..0SG/abstract

33. spiedigitallibrary Optical Engineering 63(4), 041209 — Jarzyna et al., "Photon information efficiency limits in deep-space optical communications" (2024):
34. opg.optica CLEO 2024 (Optica) — "14 Bits per Photon Detection in the Photon-Starved Regime for Deep-Space Optical Communication" (2024):
35. ui.adsabs.harvard SPIE (ADS) — "DSOC technology demonstration pre-launch validation and performance tests with the Laser Test Evaluation Station (LTES)" (2024): https://ui.adsabs.harvard.edu/abs/2024SPIE12877E..07V/abstract
36. ntrs.nasa NASA NTRS — "Testing of a Photon-Counting Optical Ground Receiver With SNSPDs and FPGA readout" (2024): https://ntrs.nasa.gov/citations/20240000366

Terrestrial Free Space Optical Communications 5

To simulate a FSOC link under real-world conditions, different tools have been investigated. After evaluating the feasibility of different simulation systems, it has been decided to use the OptiSystem by Optiwave Systems Inc. for this project [1, 2]. The user interface is very similar to Simulink.

Systems are designed with building blocks that are connected with each other. The building block libraries in OptiSystem can simulate the propagation of a laser beam in free space. Building blocks also contain different modulation techniques. In this chapter two modulation schemes were used and studied: 16-QAM and DP-QPSK. Since the expected noise on an atmospheric channel is much larger than in an optical fiber, the utilized modulation schemes have rather low order, which will likely lead to a smaller Bit-Error-Rate at a given distance. The corresponding results of the propagation limitations are documented.

5.1 16-QAM

Figure 5.1, which is a screenshot of the OptiSystem model, gives an overview over the system architecture used in the realization of an optical communication system for the 16-QAM modulation scheme.

Due to the relatively small letters of the screenshot, the components of the OptiSystems simulation system are explained in details in the following. The proposed architecture contains multiple components:

1. 16-QAM Optical transmitter: It transmits at 1550 nm with a power of 30 dBm. It is characterized with a 10, 30 and 100 cm diameter aperture.

S.-P. Chen, *Fundamentals of Free Space and Deep Space Optical Communications*,
https://doi.org/10.1007/978-3-032-35826-4_5

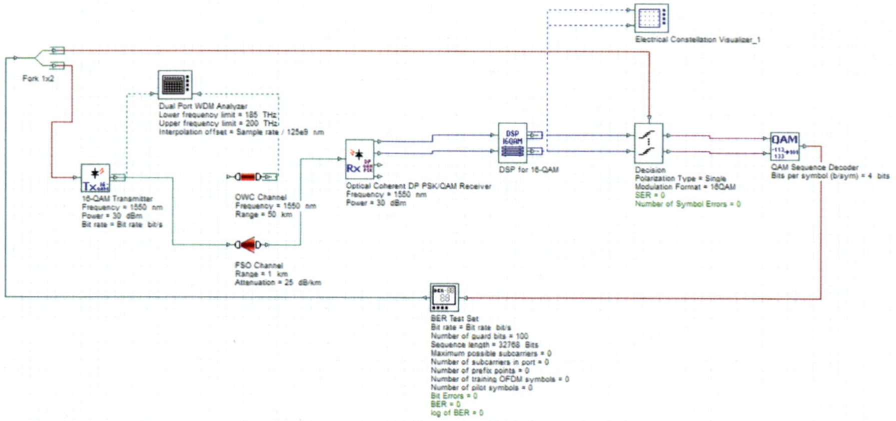

Fig. 5.1 OptiSystem 16 quadrature amplitude modulation (16-QAM) model

2. Transmission channel OWC: It models an optical wireless communication channel. It is in fact the simulation of the free space channel. In this project the attenuation characteristic is modified according to the weather conditions.
3. Optical coherent QAM receiver: It consists of a homodyne receiver which transforms incoming light into electrical power.
4. DSP for 16 QAM: It performs several functions to ensure the recovering of the incoming transmission channel after the coherent detections.
5. Decision: It processes the I and Q electrical signal channels received from the DSP component, normalizes the electrical amplitudes to the respective M-QAM grid and performs a decision on received symbols based on the preset values.
6. QAM sequence decoder: It decodes two parallel m-QAM symbol sequences to a binary signal.
7. BER test set: It calculates the bit error rate by generating large bit sequences and transmitting them to the DUT and then compares the transmitted sequence with the one received. Based on that the BER is determined.
8. Electrical constellation visualizer: Displays the in-phase and quadrature-phase electrical signals in a constellation diagram.
9. WDM analyzer: It detects, calculates and visualizes the optical power, SNR, noise etc. for each channel. Since there is only one channel this analyzer still provides a good overview on the effective attenuation behavior.

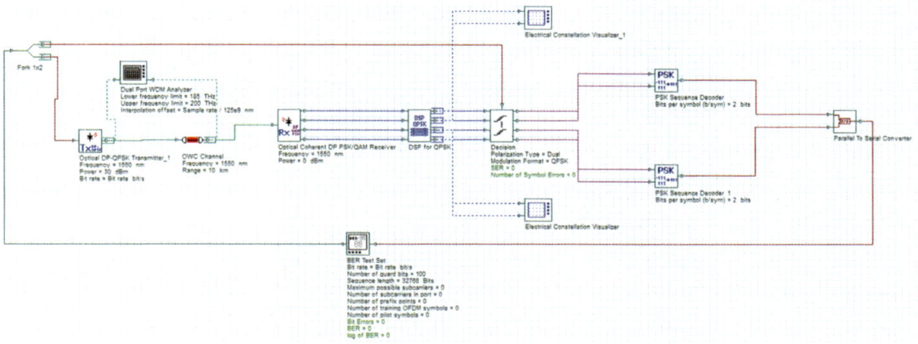

Fig. 5.2 OptiSystem dual-polarization quadrature phase-shift keying (DP-QPSK) model

5.2 DP-QPSK

Figure 5.2, which is a screenshot of the OptiSystem model, gives an overview over the system architecture used in the realization of an optical communication system for the Dual Polarization QPSK (DP-QPSK) modulation scheme.

5.3 Influence of Fog and the Transceiver Apertures

Due to the relatively small letters of the screenshot, the components of the OptiSystems simulation system are explained in details in the following. The proposed architecture contains multiple components:

1. DP QPSK Optical transmitter: It transmits at 1550 nm with a power of 15 dBm. It is characterized with a 10, 30 and 100 cm diameter aperture.
2. Transmission channel OWC: It models an optical wireless communication channel. It is in fact the simulation of the free space channel. In this project the attenuation characteristic is modified according to the weather conditions.
3. Optical coherent DP-PSK/QAM receiver: It consists of a receiver which transforms incoming light into electrical channels.
4. DSP for QPSK: It performs several functions to ensure the recovering of the incoming transmission channel after the coherent detections.
5. Decision: It separates the two polarizations and processes the I and Q electrical signal channels received from the DSP component, normalizes the electrical amplitudes to the respective PSK sequence decoders and performs a decision on received symbols based on the preset values.

6. PSK sequence decoder: It decodes two parallel PSK symbol sequences to a binary signal by using parallel to serial converter.
7. BER test set: It calculates the bit error rate by generating large bit sequences and transmitting them to the DUT and then compares the transmitted sequence with the one received. Based on that the BER is determined.
8. Electrical constellation visualizer: Displays the in-phase and quadrature-phase electrical signals in a constellation diagram.
9. WDM analyzer: It detects, calculates and visualizes the optical power, SNR, noise etc. for each channel. Since there is only one channel this analyzer still provides a good overview on the effective attenuation behavior.

Different cases have been tested using the proposed system. Based on different weather conditions (clear sky, haze, fog etc.) and different telescope apertures (10, 30, and 100 cm), the bit error rate (BER) is calculated in a distance ranging from 0 to 200 km. The diagrams are intentionally presented in linear scale to emphasize the possible transmission distances. Besides the attenuation through fog, conditions for this test case are ideal. Transceiver efficiency is set to 100%. For real-world conditions this should be set to around 80%. Also electrical noise is added to active parts of the circuit. The influence of scintillation is simulated by the optical channel. The bitrate is set to 10 Gbit/s.

Figure 5.3 shows the energy per bit over noise spectral density E_b/N_0 for terrestrial free space laser optical communications for clear weather conditions. As illustrated in Fig. 5.3

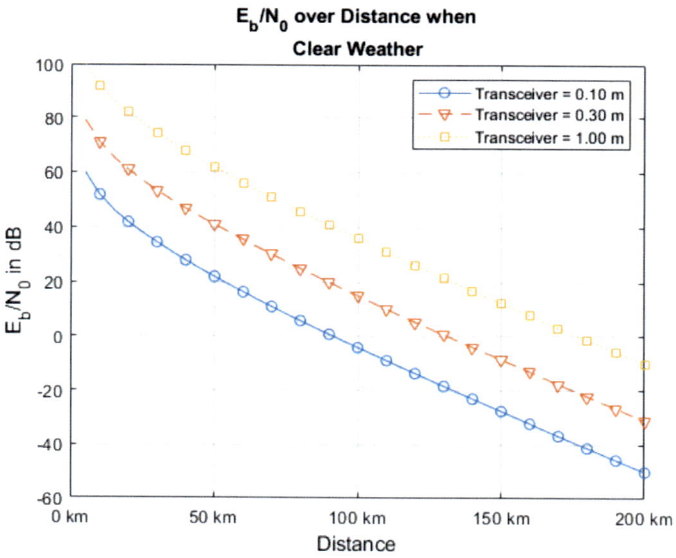

Fig. 5.3 E_b/N_0 over a distance from 0 to 200 km under clear weather conditions

E_b/N_0 decreases close to 0 in a distance of about 80 km for relatively small telescope aperture of 10 cm.

5.4 10 Cm Transceiver

Figure 5.4 shows BER versus distance in different weather conditions with a 10 cm aperture size for both the transmitter and the receiver. It is remarkable that 16-QAM and DP-QPSK are almost identical. For a low attenuation of 0.4 dB/km the transmission is possible and the BER is acceptable up to approximately 100 km. After exceeding the 100 km range the reconstruction of the transmitted signal in the receiver side is no more feasible. For higher attenuation coefficients (4 and 21 dB/km) transmission is reasonable in bad weather conditions between 2–10 km but not above this range (Table 1). For the second test the antenna diameter of the transceiver is increased from 10 to 30 cm and then 1 m.

5.5 30 Cm Transceiver

Figure 5.5 shows BER versus distance in different weather conditions with a 30 cm antenna size for both the transmitter and the receiver. It is obvious, according to the Figs. 5.4 and 5.5, that both techniques are almost identical. For a low attenuation of 0.4 dB/km transmission is possible and the BER is acceptable up to approximately 150 km. After exceeding the 150 km range, the reconstruction of the transmitted signal in the receiver side is no more feasible,

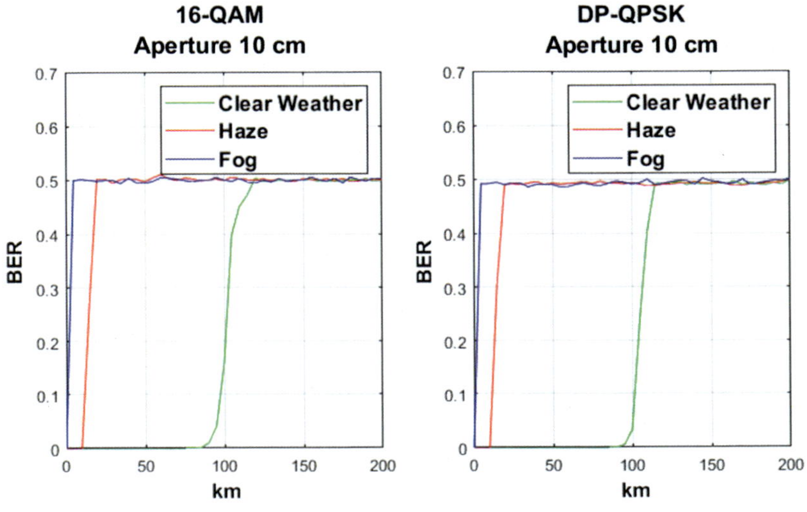

Fig. 5.4 BER for clear weather, haze, and fog versus distance; Transceiver aperture diameter 10 cm; Bit rate 10 Gbit/s

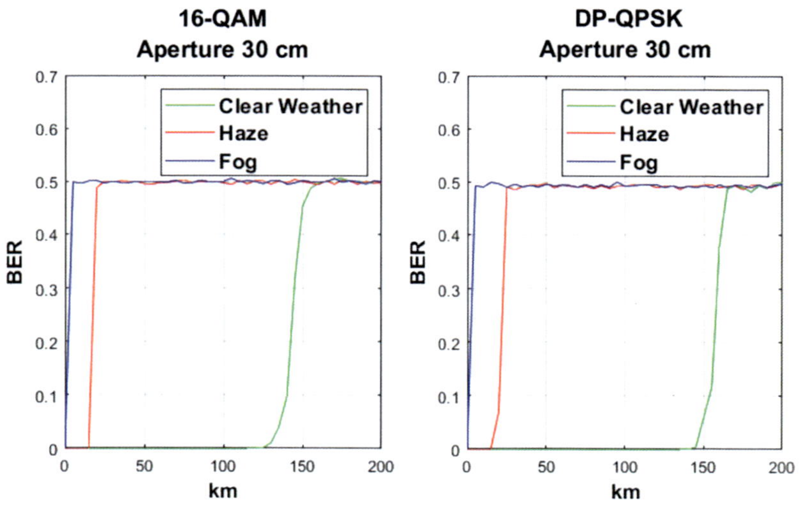

Fig. 5.5 BER for clear weather, haze, and fog versus distance; Transceiver aperture diameter 30 cm; Bit rate 10 Gbit/s

since the BER is too high. For higher attenuation coefficients (4, 21 dB/km) in bad weather conditions, transmission is achievable between 2–20 km but not above this range.

5.6 1 M Transceiver

Figure 5.6 shows BER versus distance in different weather conditions with a 1 m antenna size for both the transmitter and the receiver. Referring to the figure above, it is notable that both systems are almost identical. For a low attenuation of 0.4 dB/km the transmission is possible and the BER is acceptable up to approximately 180 km. After exceeding this range, the reconstruction of the transmitted signal in the receiver side is no more feasible, BER is too large. For higher attenuation coefficients (4, 21 dB/km) in bad weather conditions, the transmission is achievable between 2–25 km but not beyond this range.

5.7 Constellation Diagrams

The simulation results have shown that the signal quality is good until a certain distance and then the link suddenly fails. To investigate this further, constellation diagrams for DP-QPSK and 16-QAM are observed at different bitrates, distances and an attenuation of 0.4 dB/km with clear sky.

The constellation diagrams in Figs. 5.7 and 5.8 lead to a different impression. The dots are getting larger depending on the increasing distance, and some bit errors occur already at

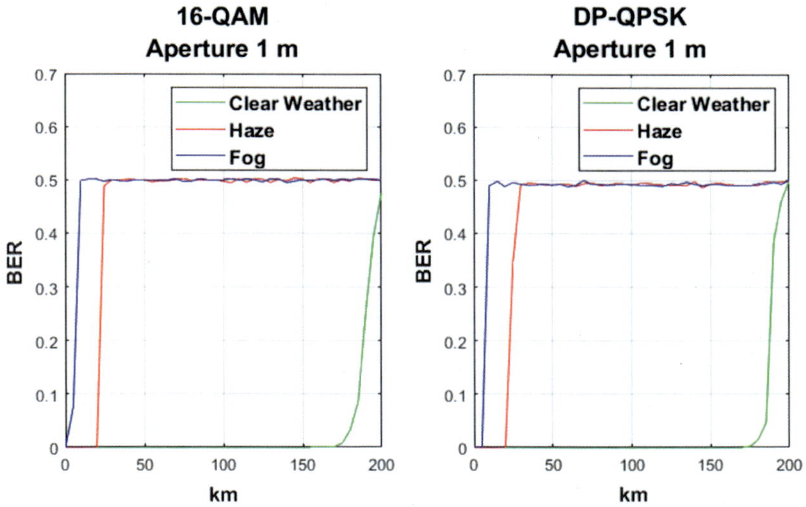

Fig. 5.6 BER for clear weather, haze, and fog versus distance; transceiver aperture diameter 1 m; Bit rate 10 Gbit/s

earlier distances, these are too small to be seen in the figures. At this point the transmitted signal quality is still good enough to demodulate the message. It is also shown clearly that a lower bitrate allows a larger transmission distance. To have a fair comparison, the signal power of the DP-QPSK is doubled to 33 dBm instead of 30 dBm, this is necessary because the constellation diagram shows only one polarization that represents half of the transmitted power.

The same results can be reproduced for the 16-QAM modulation scheme (Figs. 5.9 and 5.10). It is very interesting that at a low transmission distance of 20 km, a lower bitrate leads to a higher phase noise. At a larger distance, the signal quality for the low bitrate signal turns out to be better though.

5.8 Limitation of Higher Order Modulation Schemes M-QAM

For short distance FSOC, higher order modulation schemes up to DP-QPSK or 16-QAM can be used to achieve appropriate data rates. The TX-Laser diodes like DFB and DBR should be strictly frequency- and phase-controlled. QAM-encoder and QAM-decoder should be synchronized precisely by using Phase-Locked-Loop (PLL) techniques. The sophisticated transmitter encoder and receiver decoder circuitries are not described in this book and are assumed to be ideal.

The major disturbances are absorption, scattering, and scintillation of the atmospheric layers leading to amplitude scintillation, phase scintillation, and the limitation of transmission distance.

Fig. 5.7 Constellation diagrams DP-QPSK with 10 Gbit/s. Transceiver diameter 10 cm. Distance 20 km (top), 60 km (middle), and 80 km (bottom). Note: different scales

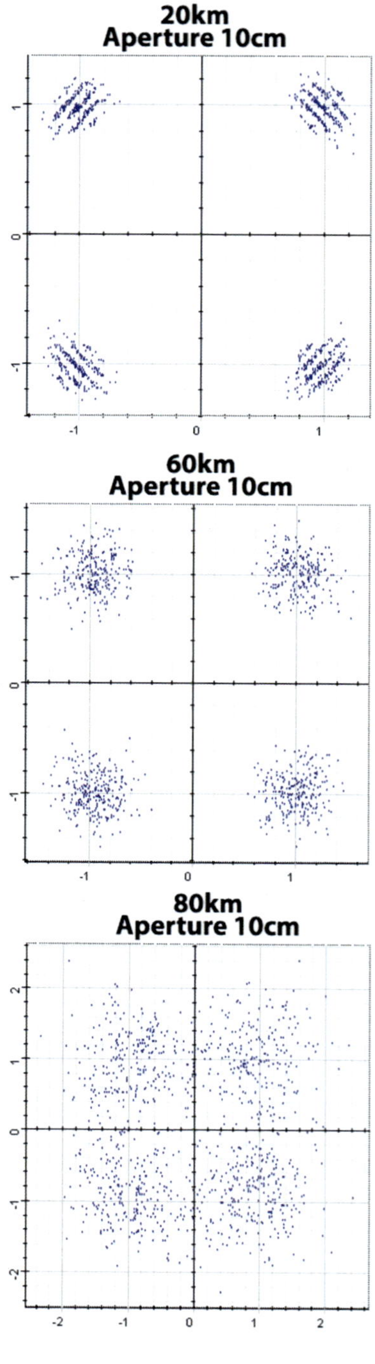

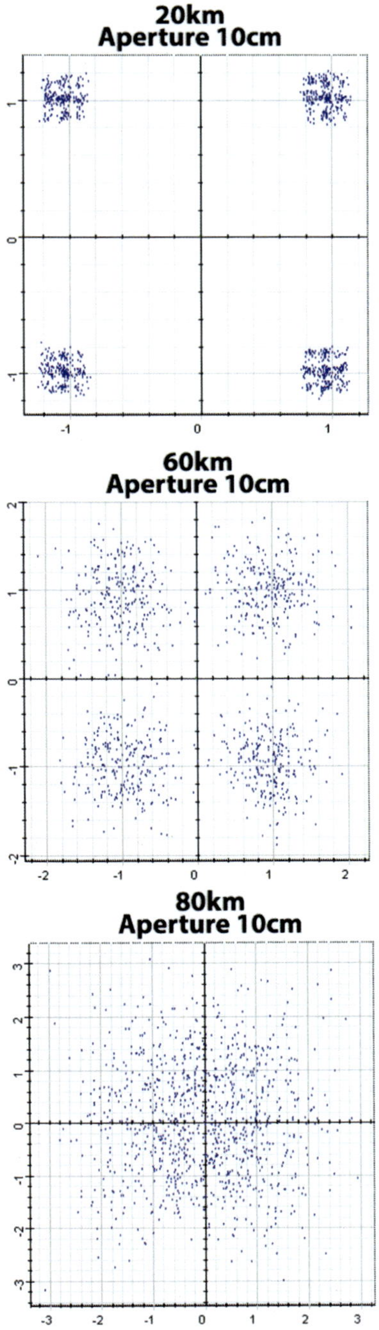

Fig. 5.8 Constellation diagrams DP-QPSK with 100 Gbit/s. Transceiver diameter 10 cm. Distance 20 km (top), 60 km (middle), and 80 km (bottom). Note: different scales

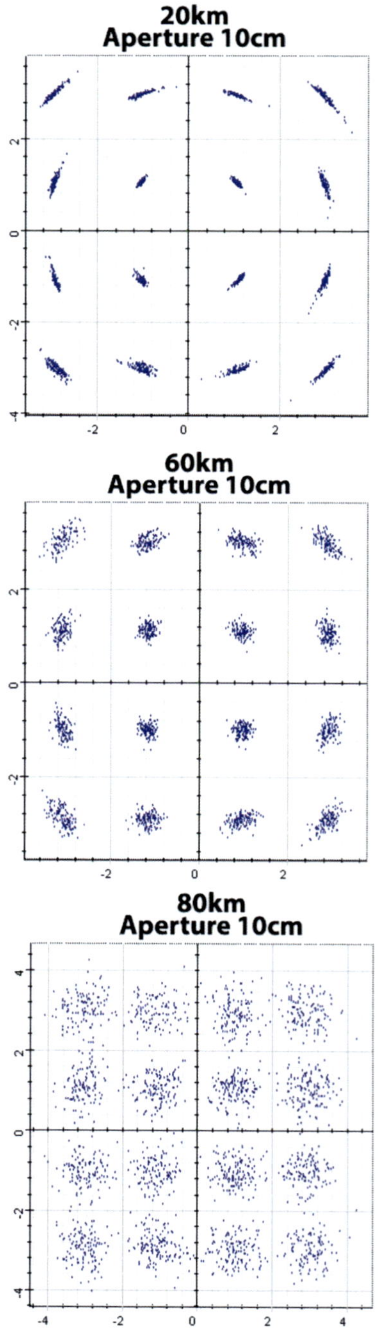

Fig. 5.9 Constellation diagrams 16-QAM with 10 Gbit/s. Transceiver diameter 10 cm. Distance 20 km (top), 60 km (middle), and 80 km (bottom). Note: different scales

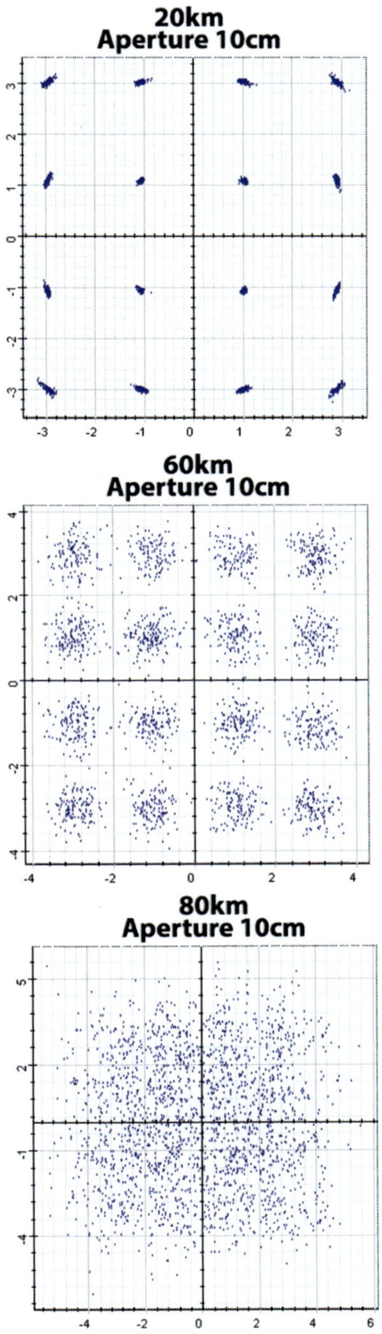

Fig. 5.10 Constellation diagrams 16-QAM with 100 Gbit/s. Transceiver diameter 10 cm. Distance 20 km (top), 60 km (middle), and 80 km (bottom). Note: different scales

5.9 Conclusions

Free space optical communications (FSOC) performances have been investigated with the simulation system OptiSystem considering the atmospheric channel, distance-dependent beam divergence, influence of weather conditions as fog and haze, turbulence, and scattering.

Interesting results with different design parameters like distances between the transmitter and receiver, modulation schemes applied, aperture diameters of the transceivers, and the influence of fog were considered and discussed. In the simulation diagrams the distance limits, with the corresponding BERs, can be seen for each modulation scheme. The transmission distance for a realistic transmission can then be chosen far enough from, or significantly smaller than, these distance limits, in order to achieve a BER of $10^{-6} - 10^{-9}$ in combination with channel coding and forward error correction. The most significant attenuation factor is beam divergence which is precisely estimated. The results also show a significant influence of non-influenceable factors. If a redundant connection at a constant bit-rate is required, the influence of atmospheric changes limits the possible distances significantly, especially the occurrence of fog requires additional planning fade margin. Besides the limitations of a single channel, the bitrates 10 Gbit/s and 100 Gbit/s can be increased by a certain factor, if coarse wavelength division multiplex (CWDM) or even dense wavelength division multiplex (DWDM) (see example [5]), which are now standard in fiber optical communication systems, can be utilized, so that Tbit/s could be possible.

References

1. Optiwave Systems Inc., "Optisystem, User Guide and Reference Manual," 2018.
2. Tobias Siegel, Shun-Ping Chen: Investigations of Free Space Optical Communications unter Real-World Atmospheric Conditions. Wireless Personal Communications, Springer Nature. https://doi.org/10.1007/s11277-020-07724-1, August 12, 2020.
3. S.-P. Chen: Investigations of free space and deep space optical communication scenarios. CEAS Space Journal, 14, 357–364 (2021). https://doi.org/10.1007/s12567-021-00388-y
4. E. D. Aboagye, S.-P. Chen: Deep space optical communications (DSOC) downlink simulation with varying PPM order. Optical and Quantum Electronics, 53, 583 (2021). https://doi.org/10.1007/s11082-021-03232-z
5. S.-P. Chen: Fundamentals of Information and Communication Technologies. Cambridge Scholars Publishing (2020).

6 Near Earth Free Space Optical Communications

6.1 Introduction

Typical laser optical communication links are discussed ranging from terrestrial, near-Earth, inter-satellite, lunar, and interplanetary scenarios [1–25]. The performance for different scenarios has been investigated by using simulation models which were successfully verified in previous projects. The simulation models consider typical mission design parameters like transmit power and operating wavelength of the optical laser transmitters, different apertures of the optical telescope, varying distances between the Earth ground station and the satellite or spacecraft terminals. In case of deep space optical communications, because of the extremely large distance and the extremely large attenuation leading to only a few numbers of received photons, photon counting receivers are indispensable [26, 27]. Atmospheric disturbances like scintillation and absorption, background noises for the uplink and downlink are considered. For relatively short distances, higher modulation schemes like M-QAM can be applied and higher bitrates can be achieved, whereas in higher orbits or at extremely large distances, appropriate pulse position modulation (PPM) orders will be necessary. Typical design parameters especially for higher orbits or large distances are those of published ESA (European Space Agency) and NASA (National Aeronautics and Space Administration) missions for GEO-orbits, lunar, and interplanetary orbits. The same method can be also used to investigate optical Inter-Satellite-Links.

For short distance free space laser optical communications (FSOC), higher modulation schemes like DP-QPSK and 16-QAM can be utilized to achieve higher throughput, e. g. 10–100 Gbit/s (details about the modulation schemes DP-QPSK and 16-QAM in the last Chap. 5).

The performance of near-Earth, lunar, and interplanetary laser optical communications is investigated by using simulation models which were successfully verified in previous projects. The simulation models consider typical mission design parameters like transmit power and operating wavelength for the optical laser transmitters, different apertures of

S.-P. Chen, *Fundamentals of Free Space and Deep Space Optical Communications*,
https://doi.org/10.1007/978-3-032-35826-4_6

the optical telescopes, varying distances between the Earth ground station and the spacecraft terminals, Geiger-mode avalanche photodiodes (APD) or superconducting nanowire single photon detection (SNSPD) receivers, atmospheric disturbances like scintillation and absorption, background noises for uplink and downlink and appropriate pulse position modulation (PPM) orders. The investigations of laser optical communications for various system parameters and different distances between the Earth ground station and the spacecraft could also help to achieve an overview of free space and deep space optical communications for different orbit constellations for future missions with laser optical communication links.

In the last decades, many projects of laser optical communications have been proposed for free space and deep space communications due to the significant advantages in comparison with classical radio frequency microwave links for space communications in terms of link performance, system size and achievable transmission bitrates. The near-Earth orbits include low Earth orbits (LEO, 100–2000 km), medium Earth orbits (MEO, 2000–36,000 km) and geostationary Earth orbits (GEO, some 36,000 km). Lunar orbits are between 350,000–410,000 km. Deep space or interplanetary links range from 0.5–3 AU (1 AU $\approx$ 150,000,000 km). Generally a distance larger than 0.01 AU, or approximately 1,500,000 km from Earth, is considered as deep space.

Many projects of laser optical communications have been proposed due to the above-mentioned significant advantages in comparison with classical radio frequency microwave links for space communications. ESA and NASA have also conducted projects like ARTEMIS, the lunar laser communication demonstration project LLCD and have planned projects like DOCS L5 and deep space optical communication links within the PSYCHE mission [1–8].

Numerous successful investigations have been performed especially for near Earth orbit optical communications, proposing different modulation schemes and system parameters [20–25]. On the other hand, deep space measurement results have been achieved from the ongoing PSYCHE mission [14].

Of course higher transmission bitrates can be achieved for relatively shorter distance of near-Earth constellations, where the signal-to-noise ratio could be designed high enough, so that higher-order modulation schemes could be applied. For deep space optical communications however, the most important requirement is to use extremely robust modulation schemes and forward error correction techniques (FEC) to enable a reliable transmission of payload data, tracking, and control data. This is because of the strongly reduced received optical power or number of the signal photons. These are the so-called photon-starved conditions. Pulse position modulation (PPM) proved to be the appropriate modulation scheme in such situations and is therefore proposed by NASA and ESA for many laser communication projects [3–8]. This is also the reason that PPM is discussed throughout this book.

Various system parameters have been investigated for laser optical communication demonstration projects, such as transmitter power, transmitter efficiency, transmitter and receiver telescope aperture gains, atmospheric loss, free space propagation loss, and receiver efficiency. In photon-starved deep space optical communications the very robust pulse position modulation scheme (PPM) in combination with Geiger-mode single photon counting avalanche photodiode (APD) receivers [26] or superconducting nanowire single photon detectors (SNSPD), are used [6, 27]. In case of the Geiger-mode APD, the single photon counting deadtime limits the slot widths, or PPM order M, and the total achievable bitrates. Besides the above-mentioned parameters, atmospheric turbulences should be considered [15, 16].

6.2 Simulation Procedures

In previous papers, the author and his co-workers have investigated (a) near-Earth short distance laser optical communications with dual polarization quadrature phase shift keying (DP-QPSK) and quadrature amplitude modulation (16-QAM) [11], (b) link budget estimations for free space laser optical communications by also considering atmospheric turbulences [10] with comparison with the experimental data of the ESA ARTEMIS project [17] and with the design parameters of the ESA DOCS L5 project [4] and (c) the impact of the varying PPM orders on the characteristics of deep space laser optical communication downlinks [13], partially with comparison with the NASA LLCD results [2, 3, 7].

The valuable measurement results achieved by ESA [5] in the ARTEMIS project for GEO satellite to Earth ground station [17], the design parameters of the DOCS L5 project [4] and the valuable experimental results achieved by NASA in the LLCD (Lunar Laser Communications Demonstration) project [2, 3, 7] were used to verify the simulation models developed at Darmstadt University of Applied Sciences in the last years. The good match between the simulation model results [10], the experimental data, and the independently ESA planned mission parameters provides the confidence to apply these models in the investigations for different scenarios in this book.

Based on these models, with some further developments, the author investigated different laser optical communication system performance characteristics depending on the varying distances during the missions and the telescope apertures. The very important noise factors reported in the above-mentioned papers were taken into account to ensure a sufficient signal-to-noise ratio and the required bitrates. These noise factors can be receiver circuit noises, direct and indirect Sun irradiances like blue sky brightness during daytime and Earth and Moon background light during nighttime. It should be noted, that the laser optical communication link can be designed more reliably, if the atmospheric turbulences characterized by the Rytov variance σ^2 are also considered. For the simulations in this book, the altitude of the ground station is assumed to be 2393 m above sea level, same as the altitude of the ideal ESA ground station at the Observatorio del Teide at Izaña in Tenerife, Spain with an average of 3700 hours of sunshine per year. The corresponding structure parameter C_n^2 of the refractive index, according to the Kolmogorov theory of turbulences and the Hufnagel-Valley Model, will then be approximately $5 \cdot 10^{-18} \mathrm{m}^{-2/3}$ corresponding to a moderate wind velocity.

The detailed descriptions of link budget simulation parameters, atmospheric scintillations and PPM can be found in [10, 11, 13] and therefore will not be explicitly repeated in this book.

In case of interplanetary missions, the orbit constellations between the Earth ground station and the spacecraft could be extremely varying, so that only certain time frames allow the data transmission with sufficient signal-to-noise ratio. In this book the reported Earth background, Moon background and blue sky brightness values for the above-mentioned missions are used for the corresponding range of distances as estimates. Also the estimated atmospheric scintillation [10] enables further analysis of the impact on the amplitude scintillation, on link budget estimations of laser optical communications. On the other hand, the channel capacity limits depend strongly on the link budget parameters like available filter bandwidth, distance, PPM orders, and signal-to-noise ratio or the ratio of received signal photons to noise photons [1, 13]. These limits represent the theoretical upper bounds, by no means the achievable bitrates. On the contrary, due to the special measures for forward error correction (FEC) like convolutional coding schemes, code rate, interleaving methods, repeat factor, and the deadtime constraints of the Geiger-mode APD single photon counting receivers, the bitrates will be much lower than the theoretical channel capacity limits.

The received power can be calculated by using the link budget formula

$$P_R(L) = P_T \cdot \eta_T \cdot g_T \cdot \eta_{Atm}(L) \cdot \eta_{FS}(L) \cdot g_R \cdot \eta_P \cdot \eta_R \tag{6.1}$$

with P_T, $P_R(L)$ as transmitted power and received power at a distance L; η_T, η_R as transmitter and receiver telescope aperture efficiencies, η_P as pointing efficiency, g_T, g_R as transmitter and receiver aperture gain; $\eta_{Atm}(L)$ as the atmospheric transmittance due to absorption loss, Rayleigh scattering (size of the scatterers $\ll \lambda$) and Mie scattering (size of the scatterers $\approx \lambda$); $\eta_{FS}(L)$ as free space propagation efficiency corresponding to the beam divergence.

By considering these link budget estimations based on the above equation (6.1), the corresponding code rates, interleaving, PPM order and the corresponding meaningful slot widths T_s can be chosen, in order to achieve the channel capacity C_{PPM} for a chosen PPM order M [18] as:

$$C_{PPM} = \frac{1}{E_\lambda \cdot \ln 2} \cdot \frac{P_R^2}{P_N \cdot \frac{1}{\ln(M)} + P_N \cdot \frac{2}{M-1} + \frac{P_R^2 \cdot M \cdot T_s}{\ln(M) \cdot E_\lambda}} \tag{6.2}$$

with P_R as received signal, P_N as noise power and E_λ as photon energy. In [13] all the components of P_N like shot noise, dark current noise, background noise and thermal noise are explained in detail and will be not repeated here again. By further considering the code rate for the forward error correction CR, the required data rate or maximum data throughput R can be estimated as:

$$R = C_{PPM} \cdot CR \tag{6.3}$$

In the NASA LLCD project [7] a reliable laser optical communication with bitrates of 10–20 Mbps for uplink and 39–622 Mbps for downlink could be successfully demonstrated by using 4-PPM for uplink and 16-PPM for downlink, both with a code rate of 1/2 and an interleaver.

Based on the received power P_R, the PPM order M and the slot width T_s including the guard slots, the number of signal photons per PPM pulse K_s can be estimated [1, 13].

In the space science community, there is an ongoing discussion about the limited aperture size due to the Fried parameter. Relatively small apertures are used for the pointing and acquisition beacon. Especially for the communication uplink for deep space either higher power or theoretically a larger aperture will be needed in order to receive enough signal photons. In both cases possibly the scintillation in the atmosphere will be increased, so it is very important to make trade-off decisions to choose the telescope apertures in an optimized way. A possible solution could be a larger aperture in combination with adaptive optics which is being investigated intensively. If this solution works out, possible further gain can be achieved. Nevertheless, in this book we will not focus on this discussion but rather on the more important issue of performance analysis. Therefore we follow the recommendations of the experts to use multiple, smaller transmitter telescope apertures for the uplink by considering the Fried parameter, with either single beam or multiple beams [19].

6.3 Near-Earth Laser Optical Communications

In Tables 6.1 and 6.2, the near-Earth laser optical communications simulation parameters are listed for uplink and downlink. Basically the parameters are chosen only as an example, but are very similar to the ESA ARTEMIS project [5] for GEO orbits. In Figs. 6.1, 6.2

Table 6.1 Simulation parameters for near-Earth laser communications, uplink [5, 17]

Near-Earth uplink	1000–40,000 km
Pulse position modulation PPM order M	4
Code rate CR	1/2
TX aperture	4 × 0.2 m
TX power	4 × 75 mW
TX laser diode wavelength	847 nm
TX efficiency	0.7
Atmospheric transmission	0.8
RX aperture efficiency with filter	0.16
RX aperture diameter	0.2 m
Earth background brightness, 5 nm filter at 38,000 km	43.6 nW/m^2 or 0.05 nW

Table 6.2 Simulation parameters for near-Earth laser communications, downlink [5, 17]

Near-earth downlink	1000–40,000 km
Pulse position modulation PPM order M	16
Code rate CR	1/2
TX aperture	0.2 m
TX power	15 mW
TX laser diode wavelength	819 nm
TX efficiency	0.7
Atmospheric transmission	0.73
RX aperture efficiency with filter	0.16
RX aperture diameter	1 m
Blue sky brightness, 3 nm filter	6 nW/m^2 or 0.75 nW

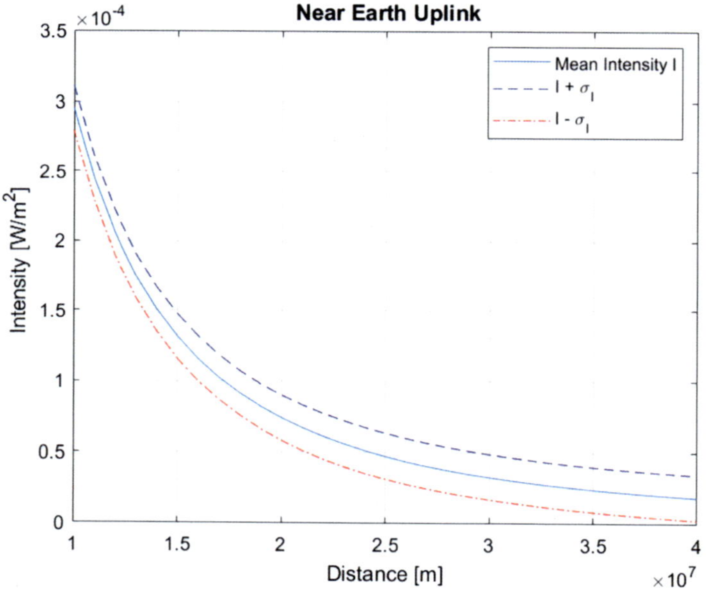

Fig. 6.1 Near-Earth uplink intensity at the receiver depending on distance based on the parameters in Table 6.1

and 6.3 the uplink characteristics, like power intensity of the laser beam, received signal photons, and achievable data throughput are displayed, whereas Figs. 6.4, 6.5 and 6.6 show the downlink characteristics. As an example the PPM orders 4 and 16 with a code rate of 1/2 are used to estimate the throughput for a communications uplink and downlink.

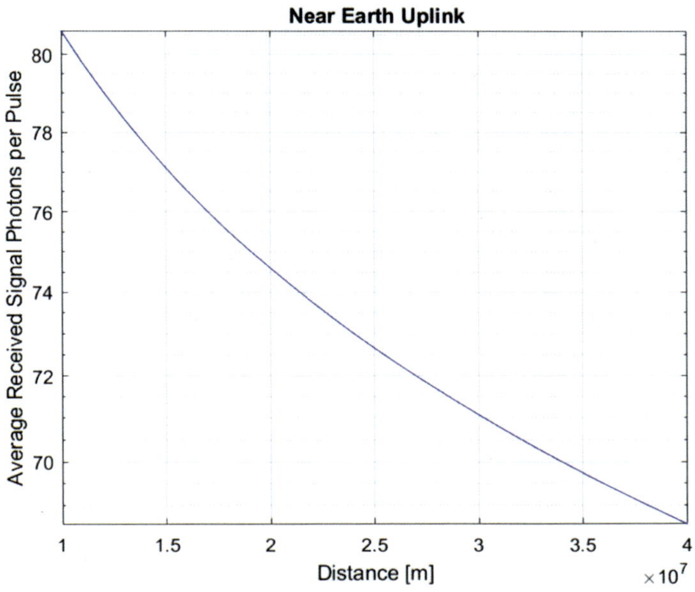

Fig. 6.2 Near-Earth uplink received signal photons per PPM pulse depending on distance based on the parameters in Table 6.1

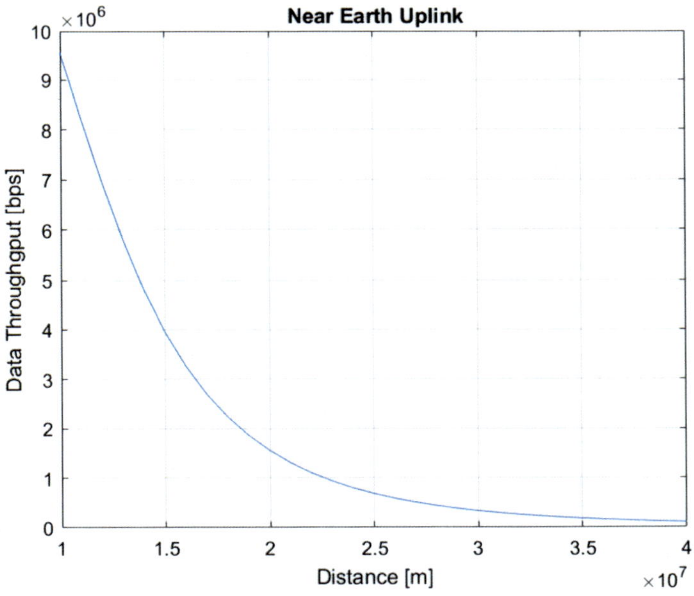

Fig. 6.3 Near-Earth uplink maximum data rate depending on distance based on the parameters in Table 6.1

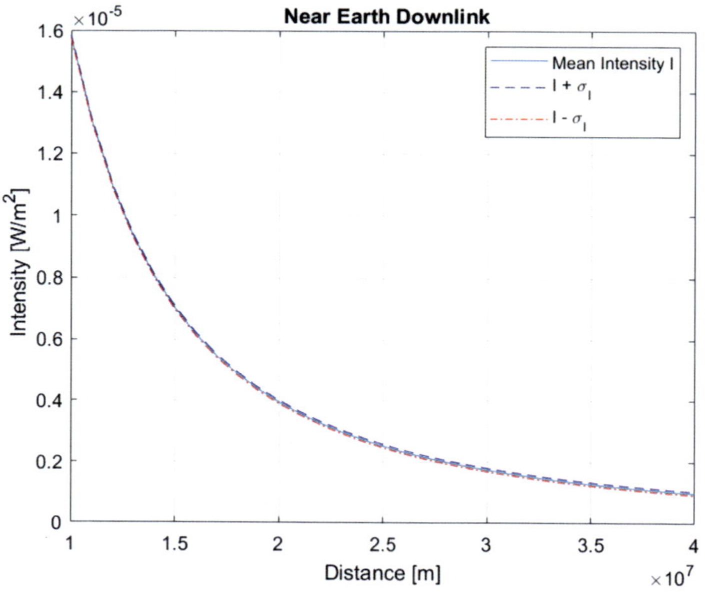

Fig. 6.4 Near-Earth downlink intensity at the receiver depending on distance based on the parameters in Table 6.2

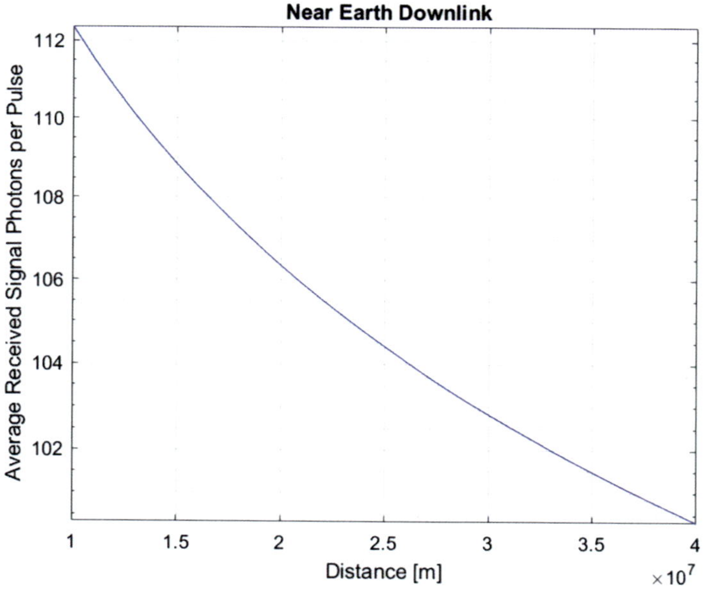

Fig. 6.5 Near-Earth downlink received signal photons per PPM pulse depending on distance based on the parameters in Table 6.2

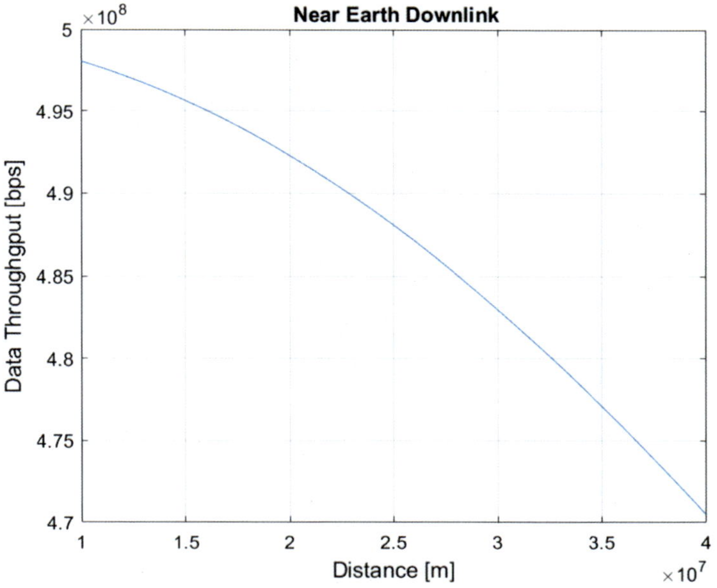

Fig. 6.6 Near-Earth downlink maximum data rate depending on distance based on the parameters in Table 6.2

Figure 6.1 shows that uplink power intensity at the receiver onboard the spacecraft, with additional deviation $\pm\sigma_I$ caused by atmospheric scintillation in the troposphere up to about 11 km, decreasing to about 0.02 mW/m^2 for a larger distance up to 40,000 km. This intensity is still much larger than the Earth background brightness of 43.6 nW/m^2. The average numbers of the signal photons per pulse are between 80–70 depending on distance (Fig. 6.2). A maximum bitrate of about 9.5–0.2 Mbit/s can be achieved correspondingly by using the link parameters from Table 6.1 (Fig. 6.3).

Figure 6.4 shows the downlink power intensity at the receiver at the ground station, with additional deviation $\pm\sigma_I$ caused by atmospheric scintillation in the troposphere up to about 11 km, decreasing to about 1 μW/m^2 for a larger distance up to 40,000 km. This intensity is still much larger than the blue sky brightness 6 nW/m^2 . The average numbers of the signal photons per pulse are between 112–100 depending on distance (Fig. 6.5). A maximum bitrate of about 500–470 Mbit/s can be achieved correspondingly by using the link parameters from Table 6.2 (Fig. 6.6).

6.4 Comparison of the Measurement Data of the Links Between GEO Satellites ARTEMIS and Earth Ground Station with the Simulation Results

ESA developed optical terminals for high data rate free space laser communication via inter-satellite (ISL) and inter-orbit links (IOL) within the frame of the Data Relay and Technology Mission (DRTM) program demonstrating for the first time that the stringent pointing, acquisition, and tracking requirements with infrared optical wavelengths is feasible [3]. For the purpose of in-orbit check of such payloads, ESA has also established an Optical Ground Station (OGS) at the Observatorio del Teide at Izaña, Tenerife, Spain. Tenerife has the ideal condition with extremely low atmospheric loss and scintillation. The OGS communicates with the optical laser system OPALE (Optical Payload Laser Experiment) onboard ARTEMIS. The measurement campaign was conducted from 2003 to 2004 and valuable measurement data were collected. By using these measurement data, the link budget calculation can be verified and also the scintillation estimation can be validated.

The basic downlink and uplink parameters are listed in Tables 6.3 and 6.4. The simulation results for uplink and downlink, with consideration of the scintillation effect, are compared

Table 6.3 Downlink ARTEMIS OPALE to OGS

ARTEMIS OPALE—OGS	38,018 km
TX power	15 mW
TX laser diode wavelength	815 nm
TX efficiency	0.7
TX aperture	10 cm or 20 cm
Atmospheric transmission	0.73
RX aperture efficiency with filter	0.16
RX aperture diameter	1 m
Blue sky brightness, 3 nm filter	6 nW/m^2

Table 6.4 Uplink OGS to ARTEMIS OPALE

OGS—ARTEMIS OPALE	38,018 km
TX power 4 ×	75 mW
TX laser diode wavelength	847 nm
TX efficiency	0.7
TX aperture 4 ×	10 cm
Atmospheric transmission	0.8
RX aperture efficiency with filter	0.16
RX aperture diameter	0.1 m
Earth background brightness, 5 nm filter	43.6 nW/m^2

Table 6.5 Comparison of simulation (DL, UL) with measurement data [17]

	Mean irradiance I	Scintillation σ_I
Artemis measurement DL	33.76 nW/m^2	2.29 nW/m^2
Simulation DL	32.00 nW/m^2	1.30 nW/m^2
Artemis measurement UL	122.84 nW/m^2	43.81 nW/m^2
Simulation UL	103.69 nW/m^2	29.22 nW/m^2

with those in Table 6.5. For the simulation, first the link budget for irradiance I, which represents the mean value of the stochastic intensity variations due to the scintillation effect and received power P, is estimated. Then the index of the refractive structure parameter C_n^2, which will be influenced by wind velocity, temperature, and atmospheric pressure jointly, is calculated. At the end, the Rytov variance σ_I^2 as stochastic variation of the intensity, which corresponds to the standard variation of the stochastic process with a sampling period of 20 ms, will be derived from C_n^2. The results were combined and compared with the ARTEMIS measurement data. C_n^2 is chosen to be $5 \cdot 10^{-18} m^{-2/3}$ for both downlink and uplink, to achieve the best match, since the site of the OGS in Tenerife is on top of the mountain with a height of h = 2393 m above the sea level and the measurements were performed during different daytimes. Since there are no contemporaneous weather measurement data like wind velocity, atmospheric pressure, and temperature available, we could not consider the possibly changing weather parameters dynamically and instead have chosen only the above mentioned effective C_n^2 parameter with which all simulation results match pretty well with the measurement data. The comparison shows that the simulation method can be used for the further link budget investigations and project design with sufficient accuracy.

The simulation gives the mean irradiance based on spherical beam propagation and/or Gaussian beam propagation with identical results. The irradiance varies with time, described by the Rytov variance σ_I^2. For the simulation, the time series have been matched to the measurement data scenario. The irradiance variation is obtained from the standard deviation σ_I and is simulated with Gaussian distributed random numbers for subsequent data points.

The measurement of the downlink laser transmission experiment started on May 6, 2003 at 22:15 h and stopped after 914 s. During the measurement of the uplink laser transmission experiment starting on May 24, 2003 at 17:00 h, after the link acquisition, the transmission bit error rate (BER) analysis started 278 seconds later. The transmitter switched the uplink lasers gradually from 4 lasers to 3 lasers 898 seconds after the start, then to 2 lasers after 1348 seconds, and finally to 1 laser after 1588 seconds. The measurement was stopped after 1798 seconds. The differences between simulation and measurement results should be further analyzed. Figure 6.7 shows the measured downlink, Fig. 6.8 shows the simulated downlink, Fig. 6.9 shows the measured uplink, and Fig. 6.10 shows the simulated uplink. As discussed before, due to the missing simultaneous weather measurement data which could lead to significant changes of the C_n^2 values, we assumed an effective constant C_n^2 value

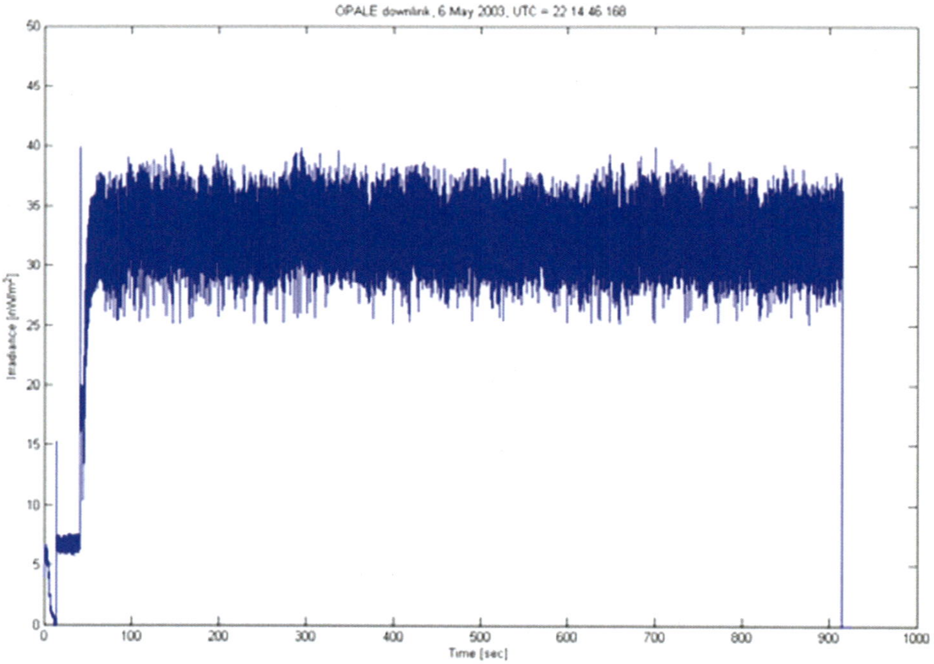

Fig. 6.7 Measurement results for Artemis downlink [4]. Irradiance (nW/m^2) versus Time (s)

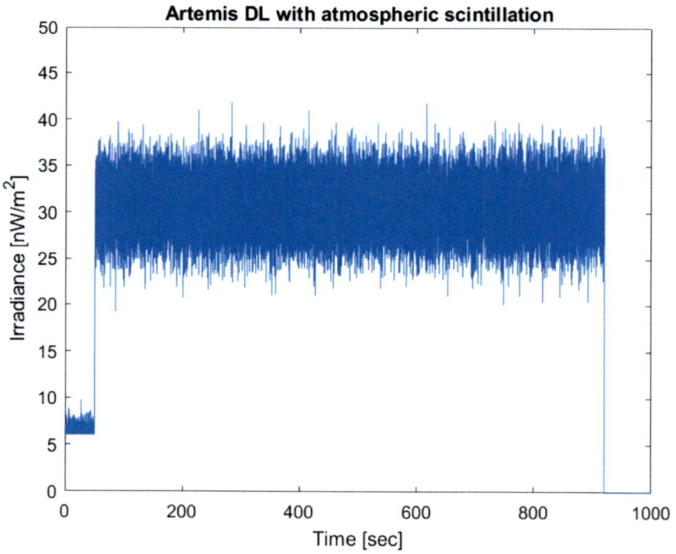

Fig. 6.8 Simulation results for ARTEMIS downlink. Irradiance (nW/m^2) versus Time (s)

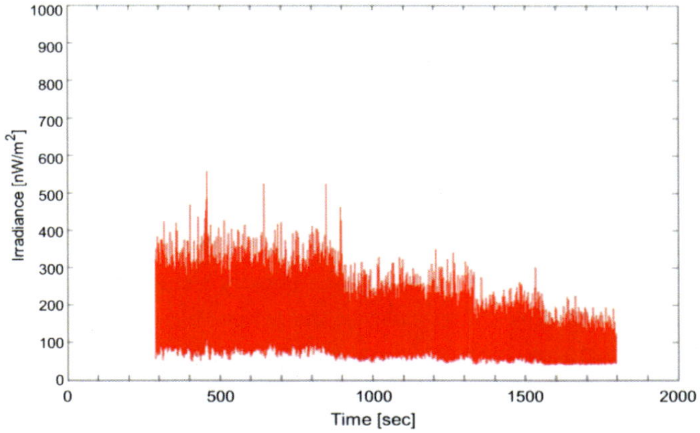

Fig. 6.9 Measurement results for Artemis uplink [4]. Irradiance (nW/m^2) versus Time (s)

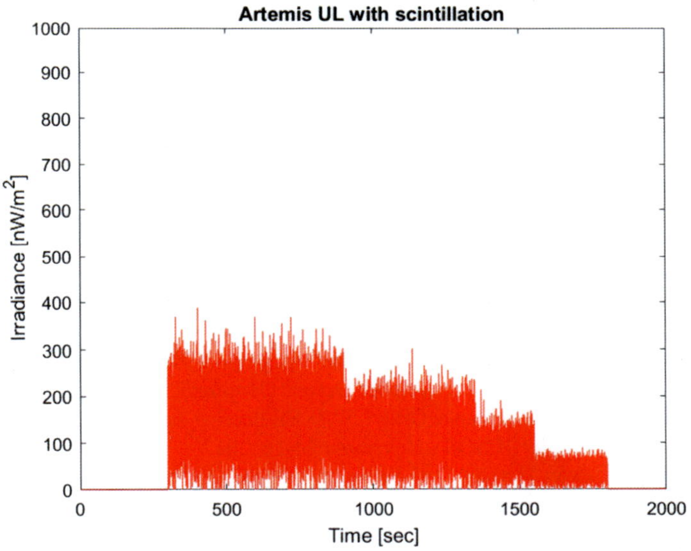

Fig. 6.10 Simulation results for ARTEMIS uplink. Irradiance (nW/m^2) versus Time (s)

for the complete measurement duration. This explains also the rather smaller difference between the simulation results and measurement results at the beginning with 4 laser beams compared with the case with only 1 laser beam at the end, possibly due to the temperature change and/or wind velocity changes during the measurement.

6.5 Conclusions

Simulation results by using previously verified algorithms for near-Earth, lunar and interplanetary laser optical communications are presented in this chapter. Different realistic system parameters are considered like uplink and downlink transmit power, telescope apertures, different distances, atmospheric scintillations, blue sky brightness, Earth and Moon background noise as well as receiver noise factors. Generally, by using the assumed mission parameters, the required received power or power density levels can be ensured. The results show the uplink and downlink characteristics, the achievable data rates, give an overview and could possibly help to verify future mission planning and measurement data, and further improve the performance by modifying uplink and downlink parameters.

References

1. H. Hemmati, A. Biswas, I. B. Djordjevic, "Deep Space Optical Communications: Future Perspectives and Applications", Proceedings of IEEE, vol. 99, No. 11, November 2011.
2. M. Sans and Z. Sodnik, I. Zayer and R. Daddato, "Design of the ESA Optical Ground Station for Participation in LLCD", Proc. International Conference on Space Optical Systems and Applications (ICSOS) 2012, 3-1, Ajaccio, Corsica, France, October 9-12 (2012).
3. Z. Sodnik, H. Smit, M. Sans, D. Giggenbach, P. Becker, R. Mata-Calvo, C. Fuchs, I. Zayer, M. Lanucara, K.-J. Schulz, J. Widmer, F. Arnold, M. Mosberger, A. Alonso, I. Montilla, "Results from a Lunar Laser Communication Experiment between NASA's LADEE Satellite and ESA's Optical Ground Station", Proc. International Conference on Space Optical Systems and Applications (ICSOS) 2014, S2-1, Kobe, Japan, May 7-9 (2014).
4. Zoran Sodnik, Clemens Heese, Pantelis-Daniel Arapoglou, Klaus-Jürgen Schulz, Igor Zayer, Robert Daddato, Stefan Kraft, "Deep-Space Optical Communication System (DOCS) for ESA's Space Weather Mission to Lagrange Orbit L5". IEEE International Conference on Space Optical Systems and Applications (ICSOS), November 14-16, 2017.
5. Z. Sodnik and R. Czichy, "Design Data Summary of the ESA Optical Ground Station for In-Orbit Check-Out of Laser Communication Payloads and for the Observation and Registration of Space Debris". ESA Specifications for the Artemis OGS OPALE project. Ref.: TEC-MMO/267/ZS. May 2, 2005.
6. J. Mendenhall, L. Candell, P. Hopman, G. Zogbi, D. Boroson, D. Caplan, C. Digenis, D. Hearn, and R. Shoup, "Design of an Optical Photon Counting Array Receiver System for Deep-Space Communications", Proceedings of IEEE, vol. 95, No. 10, October 2007.
7. D. Cornwell, D. Boroson, B. Robinson, D. Burianek, D. Murphy, F. Khatri, "The Lunar Laser Communication Demonstration (LLCD): NASA's First High-Rate, Two-Way Space Laser Communications Demonstration". Presentation to the JHU Aerospace Affinity Group. June 11, 2014.
8. A. Biswas, "NASA's Deep Space Optical Communications – An Update". OSA Laser Congress, Vienna, Austria. October 3, 2019.
9. Tobias Siegel, Shun-Ping Chen: Investigations of Free Space Optical Communications. Wireless Personal Communications (2021) 116, 475–490. https://doi.org/10.1007/s11277-020-07724-1.
10. Shun-Ping Chen: Performance analysis of near earth, lunar and interplanetary optical communication links. Optical and Quantum Electronics (2022) 54. https://doi.org/10.1007/s11082-022-03987-z

11. S.-P. Chen, "Investigations of free space and deep space optical communication scenarios", CEAS Space Journal, Springer Nature, September 13, 2021. https://doi.org/10.1007/s12567-021-00388-y.
12. P. Thielke, F. Schüßler, T. Stieglitz, S.-P. Chen: FSOC and DSOC. Internal Research Project Report. Institute of Communication Technologies. Darmstadt University of Applied Sciences. 26.06.2024
13. Emmanuel Domfeh Aboagye, Shun-Ping Chen: Deep space optical communications (DSOC) downlink simulation with varying PPM order. Optical and Quantum Electronics (2021) 53. https://doi.org/10.1007/s11082-021-03232-z
14. NASA, "NASA's Optical Comms Demo Transmits Data Over 140 Million Miles". (2024), URL: https://www.nasa.gov/missions/psyche-mission/nasasoptical-comms-demo-transmits-data-over-140-million-miles/ (Retrieved on May 15, 2024).
15. Larry C. Andrews, Ronald L. Phillips, Cynthia Y. Hopen, "Laser Beam Propagation through Random Media", SPIE Press, Bellingham, Washington USA, 2005.
16. Larry C. Andrews, Ronald L. Phillips, Cynthia Y. Hopen, "Laser Beam Scintillation with Applications", SPIE Press, Bellingham, Washington USA, 2001.
17. Internal ESA Artemis OGS OPALE Measurement Results. 2003-2004.
18. B. Moision, William Farr, "Range Dependance of the Optical Communications Channel", IPN Progress Report 42-199, November 15, 2014.
19. R. M. Calvo, P. Becker, D. Giggenbach, F. Moll, M. Schwarzer, M. Hinz, Z. Sodnik, "Transmitter diversity verification on ARTEMIS geostationary satellite"', SPIE Photonics West, 2014.
20. H. Zech, Ph. Biller, F. Heine, M. Motzigemba, "Optical Intersatellite Links for Navigation Constellations", ICSO 2018 International Conference on Space Optics. Chania, Greece. 9–12 October 2018
21. S. Chaudhary, X. Tang, A. Sharma, B. Lin, X. Wei, A. Parmar, "A cost-efective 100 Gbps SAC-OCDMA–PDM based inter-satellite communication link". Optical and Quantum Electronics (2019) 51:148 https://doi.org/10.1007/s11082-019-1864-2.
22. D. Dubey, Y. K. Prajapati, R. Tripathi, "Error performance analysis of PPM-and FSK-based hybrid modulation scheme for FSO satellite downlink"'. Optical and Quantum Electronics (2020) 52:286 https://doi.org/10.1007/s11082-020-02404-7.
23. M. Singh, J. Malhotra, "A high-speed long-haul wavelength division multiplexing–based inter-satellite optical wireless communication link using spectral-efficient 2-D orthogonal modulation scheme". International Journal of Communication Systems. 2020;33:e4293. John Wiley & Sons, Ltd. https://doi.org/10.1002/dac.4293.
24. M. Singh, J. Malhotra, "Modeling and Performance Analysis of 400 Gbps CO-OFDM Based Inter-satellite Optical Wireless Communication (IsOWC) System Incorporating Polarization Division Multiplexing with Enhanced Detection". Wireless Personal Communications (2020) 111:495–511. https://doi.org/10.1007/s11277-019-06870-5.
25. A. Viswanath, V. K. Jain, S. Kar, "Reduction in transmitter power requirement for earth-to-satellite and satellite-to-earth free space optical links with spatial diversity". Opt Quant Electron (2018) 50:418 https://doi.org/10.1007/s11082-017-1257-3.
26. M. Srinivasan, K. S. Andrews, W. H. Farr, A. Wong, "Photon Counting Detector Array Algorithms for Deep Space Optical Communications", Proc. of SPIE Vol. 9739, 97390X, https://doi.org/10.1117/12.2217971.
27. C. M. Natarajan, M. G. Tanner, R. H. Hadfield, "Superconducting nanowire single-photon detectors: physics and applications", Superconductor Science and Technologies, 25 (2012), https://doi.org/10.1088/0953-2048/25/6/063001.

Pulse Position Modulation for Deep Space Optical Communications 7

During the course of a typical deep space mission like the Earth-Mars mission, there exist a wide range of operating points, due to the different changes in geometry that consequently cause different link budgets in terms of received signal and noise power. These changes include distance range, Sun-Earth-Probe angle (SEP), zenith angle, and atmospheric conditions. The different operating points, with different losses (background noise, pointing losses, and atmospheric losses), lead to different capacities and data rates over the course of a typical deep space mission. Consequently, different engineering parameters are adjusted and optimized to combat some of these varying losses in order to get acceptable data rates and bit error probabilities. This is the reason to analyze and simulate various operating conditions that occur with the varying spatial orbital time periods, like resulting received signal power level, noise power level, capacity, data rates, and bit error probabilities. This chapter describes detailed results of simulations of typical deep space optical communication link operations.

7.1 Introduction

There has been a paradigm shift in the use of radio communications to optical communications in the field of deep space communications during the last decades. The reason for this increasing interest in applications of optical communications in deep space is due to its unique attributes such as large bandwidth, license free spectrum, high data rate, efficient power utilization, and low mass requirements.

Deep space optical communications (DSOC) involve a communication link that focuses mainly on the transmission between the distant planetary objects that are millions of kilometers apart from the Earth. One famous example is the communication link between the

S.-P. Chen, *Fundamentals of Free Space and Deep Space Optical Communications*,
https://doi.org/10.1007/978-3-032-35826-4_7

Earth and a spacecraft surrounding another planet such as Mars. The distance range between these planetary objects may be described in astronomical units (AU).

This book presents detailed results of simulations conducted in a typical DSOC Earth-Mars link operation. During the course of a typical Earth-Mars mission, there exist a wide range of operating points due to the different changes in geometry that consequently lead to different Earth-Mars link budgets in terms of received signal and noise power. Examples of these parameters during the course of a Earth-Mars mission include distance range, Sun-Earth-Probe angle (SEP angle), zenith angle and atmospheric conditions. These different operating points can cause different losses (background noise, pointing losses, and atmospheric losses) leading to different channel capacities and data rates over the course of a typical Earth-Mars mission. Consequently, different parameters are adjusted to combat some of these varying losses in order to get acceptable data rates and bit error probabilities. Thus, over the course of a typical Earth-Mars mission, there will be varying noise power levels and received signal power levels due to the varying orbital time periods. For example, there will be orbital time instances when the Sun will be between Mars and Earth leading to higher level of noise due to the small SEP angle and also a higher distance range between Mars and Earth (the largest distance between Earth and Mars is about 225 million km) [1–3]. Conversely, in other orbital time instances, there will be situations where the Sun will not be between Mars and Earth. This causes lower level of noise and also there may be lower distance between Mars and Earth (the closest distance of Earth to Mars is 56 million km). The ratio of the distance ranges of these two distinct positions over an orbital period (2.2 years) is approximately 4. All these factors lead to different channel capacities, data rates and bit error probabilities in these different conditions. Different engineering input parameters are adjusted accordingly to maintain or improve the data rates, and bit error rates as the spatial orbital time period varies.

This provides a good basis to undertake analyses and simulations of the various operating conditions that occur with the varying spatial orbital time periods on the resulting received signal power level, noise power level, capacity, data rates and bit error probabilities.

In the following sections of this book, a deep space optical communication system is modeled and simulated by using Matlab. The impact of optimization of the number of received signal photons and noise photons, channel capacity, data rates and bit error rates through tuning and adjustment of the various parameters such as PPM order, laser transmitter aperture size, receiver aperture size and laser transmit power will also be analyzed in this book. A photon counting model in a non-coherent detection system setup for intensity modulation is used in these simulations to determine the photon counts at the receiver side to consequently recover the data transmitted. This is because non-coherent detection is the best method to achieve energy efficiency which is indispensable in deep space. Furthermore, Pulse-Position Modulation (PPM) is used in the downlink simulations due to higher energy efficiency with low duty cycles compared with On-Off-Keying (OOK) in deep space when transmitting from Mars to Earth [1, 4–7].

7.2 DSOC Systems

The deep space optical communication DSOC system model consists of the laser transmitter (TX), TX aperture gain, DSOC channel, receiver (RX) aperture gain and the APD receiver (RX). Fig. 7.1 shows a block diagram of the model. The DSOC channel comprises space loss, pointing losses, atmospheric effects and optical background of the link between the transmitters and receivers stationed on Earth and Mars [1].

In M-PPM, each symbol interval is divided into M time slots and a non-zero optical pulse is placed in one of these time slots while other slots are kept vacant. Moreover, for the purpose of synchronization additional slots called synchronization slots are added. For long distance or deep space communications, the M-PPM scheme is widely used because it provides a high peak-to-average power ratio (PAPR) that improves its average-power efficiency [8]. Since k bits are sent per symbol, the average number of signal counts required per bit is divided by k, which makes it more energy efficient than modulation for $k > 2$. However, the M-PPM scheme has poor bandwidth efficiency at higher values of M. Downlink simulations were done for PPM orders ranging from 2 to 4096; nevertheless 4 to 256 is typically used in the deep space mission for Mars [1, 7].

The average power for M-PPM can be calculated as

$$P_{avg} = \frac{(P_{max} + P_{min} \cdot (M - 1))}{M} \tag{7.1}$$

The signal slot width Ts is given by the equation depending on the parameter BR

$$T_s = \frac{1}{PRR} \tag{7.2}$$

The pulse repetition rate PRR required for a system with PPM order M, bit rate BR and k bits per symbol is

$$PRR = \frac{M \cdot BR}{k} \tag{7.3}$$

The free space loss FSL is dependent on the distance (range) d between the transmitter and receiver. It is

Fig. 7.1 Deep space optical communications (DSOC)-blockdiagram

$$FSL = 20 \cdot \log\left(\frac{4\pi d}{\lambda}\right) \tag{7.4}$$

The transmitter aperture gain (TX gain) and receiver aperture gain (RX gain) are dependent on the telescope aperture diameter D in meters, the optical wavelength λ in meters and the optical efficiency η of the lens. The gains are [9]:

$$G = \eta \cdot \left(\frac{\pi D}{\lambda}\right)^2 \tag{7.5}$$

The background radiation interferes with the received signal. Objects producing or reflecting light, such as the Sun, the stars, and Earth, may interfere with the received signal and cause background noise. Received background power P_{back} depends on the background irradiance, effective receiver area, receiver field of view, optical filter bandwidth, and the optical efficiency and it is modeled as

$$P_{back} = H_b \cdot A_{rec} \cdot \eta \cdot \gamma_{FOV} \cdot \Delta\lambda_{nr} \tag{7.6}$$

where H_b is the background irradiation energy density (W/m^2/sr/μm), A_{rec} is the effective receive area (m^2), γ_{FOV} is the receiver field of view measured in steradians (sr), $\Delta\lambda_{nr}$ is the optical filter bandwidth (μm) and η is the optical efficiency of the receiver (scalar). The receiver field of view is

$$\gamma_{FOV} = \frac{\pi}{4} \cdot \theta_{FOV}^2 \tag{7.7}$$

where θ_{FOV} represents the planar angular detector field of view (radians) and depends on the telescope aperture. The value of sky radiance, H_b ranges from 0.007–0.015 W/(cm^{2}srμm) depending on the zenith angle and the time of daylight [1].

Adverse atmospheric effects impact on the performance of the optical link. Consequently, during the design of the ground station receiver system, measures are factored into the process in order to reduce the effects of adverse atmospheric conditions. These atmospheric effects include: atmospheric scintillation, atmospheric irradiance fluctuations and atmospheric turbulence induced angle of arrival jitter fluctuations, atmospheric transmittance and atmospheric "seeing". One method is the use of large effective receiver aperture diameter during the design of the DSOC downlink system in order to collect more photons. The side-effect associated with such design is the increased impact of background noise through sky radiance. This is factored into the background noise calculation through the value specified for sky radiance. The negative side-effect of using large receiver aperture diameter is also minimized by restricting the detector field of view (FOV) in order to reject excessive background noise from sky radiance. Moreover, special spectral filters are used to reduce the collected sky radiance or background flux. Another technique for reducing the sky radiance is the use of a form of a distributed optical antenna array at the ground station receiver.

Furthermore, the negative impact of adverse atmospheric turbulence is reduced by employing large aperture averaging for the receiver with diameters ranging from 5 m to 10 m whilst the Fried parameter r_0 is 0.05 m. This provides high D_{rx}/r_0 that reduces the impacts of atmospheric scintillation, irradiance fluctuations, "seeing effects" and induced angle of arrival jitter fluctuations according to pages 184–207 of literature [1]. An atmospheric scintillation penalty of 0.1 dB is assumed in the link budget.

Atmospheric transmission loss is

$$T_{atm} = e^{M_r} \cdot T_{zenith} \tag{7.8}$$

where T_{zenith} is the atmospheric zenith transmittance and M_r is the relative air mass.

Other factors like the choice of location of the ground station and the time (day or night) for receiving data on the downlink of the DSOC system is very vital in mitigating against some of the adverse atmospheric impacts. Pages 169–184 of literature [1] subtitled "3.3 Atmospheric Issues on Ground Telescope Site Selection for an Optical Deep Space Network" provide an overview of mitigating some of the impacts of the atmospheric turbulence effects. For example, the altitude of the location of the ground station and the zenith angle during the downlink designs. Thus, ground station locations are carefully chosen in order to limit the impact of atmospheric effects and losses as low as possible. Finally, the choice of 1.55μm as the optical wavelength for the downlink of the DSOC system help to further mitigate against atmospheric losses such as absorption and scattering of the light beam due to gas molecules and aerosols present in the atmosphere.

The incident light into the APD is converted into electrical signals proportional to the optical power and the responsivity is

$$R = \frac{q \cdot \eta}{h \cdot f} \tag{7.9}$$

where q is the electron charge, η the quantum efficiency, h the Planck's constant, and f is the optical frequency.

A stream of discrete photons is emitted from a laser transmitter and collected by the APD receiver. In order to build a realistic APD receiver model, noise originating within the diode is considered. Shot noise depends on the average current from different sources and is modeled by

$$\sigma_{sn}^2 = 2 \cdot q \cdot R \cdot P_r \cdot G^2 \cdot F \cdot B_w \tag{7.10}$$

where G is the optical amplifier gain, Pr is the incident received optical power , B_w is the electrical bandwidth, F is the excess noise factor of the APD and is

$$F = k_{eff} \cdot G + (1 - k_{eff}) \cdot (2 - 1/G) \tag{7.11}$$

where k_{eff} is the ratio of the hole and electron ionization coefficients.

Background shot noise produced by a similar process as signal shot noise is

$$\sigma_{bg}^2 = 2 \cdot q \cdot R \cdot P_{back} \cdot G^2 \cdot F \cdot B_w \tag{7.12}$$

where P_{back} represents the background noise power.

The dark current shot noise is

$$\sigma_{dc}^2 = 2 \cdot q \cdot F \cdot G^2 \cdot I_b \cdot B_w + 2 \cdot q \cdot I_s \cdot B_w \tag{7.13}$$

where I_b is the bulk leakage current which becomes amplified and I_s is the surface leakage current which does not pass through the avalanche region.

The thermal noise is

$$\sigma_{th}^2 = \frac{4 \cdot k_B \cdot T_r \cdot F \cdot B_w}{R_L} \tag{7.14}$$

where R_L is the load resistance, T_r the electronic system noise temperature, and k_B the Boltzmann constant. The temperature T_r is the equivalent temperature of the loss and is usually the physical temperature of the load resistor.

Aggregating all currents and noise current sources, the mean and variance of the total output from the receiver can be derived. Two scenarios are considered: firstly when a signal pulse is received and secondly when there is no pulse received between the pulses. The mean output current from the receiver for both cases is

$$\mu_1 = R \cdot G \cdot P_{r1} + R \cdot G \cdot P_b + G \cdot I_b + I_s \tag{7.15}$$

$$\mu_0 = R \cdot G \cdot P_{r0} + R \cdot G \cdot P_b + G \cdot I_b + I_s \tag{7.16}$$

where the index "1" denotes a received pulse and a "0" denotes the absence of a pulse. The variance of the output current is

$$\sigma_1^2 = \sigma_{sn1}^2 + \sigma_{bg}^2 + \sigma_{dc}^2 + \sigma_{th}^2 \tag{7.17}$$

$$\sigma_0^2 = \sigma_{sn0}^2 + \sigma_{bg}^2 + \sigma_{dc}^2 + \sigma_{th}^2 \tag{7.18}$$

The signal to noise ratio (SNR) is determined and the bit error rate is calculated from these variances and mean output currents [4, 5].

The capacity of the DSOC system was determined from the equation given by [10, 11] in page 93 of [1] as

$$C = (\log_2 e)\frac{\lambda}{M} \cdot \left[(1 + \frac{1}{\rho}) \cdot \ln(1 + \rho) - (1 + \frac{M}{\rho}) \cdot \ln(1 + \frac{\rho}{M})\right] \tag{7.19}$$

The data rate, R is

$$R = \frac{C}{M} \cdot \log_2 M \tag{7.20}$$

where C is the channel capacity and M is the PPM-order.

Data throughput will be in combination with the code rate CR

$$Throughput = R \cdot CR \tag{7.21}$$

Simulations of the received signal photons, noise photons, channel capacity, data rate, and data throughput are also computed accordingly. All equations are obtained from reference [1]. Table 2.2 gives the list of parameters and values used for the simulation [2, 3]. Synchronization slots are not considered in the simulations. Bit SNR is explained as slot SNR divided by the number of bits in a PPM Symbol derived in pages 258–259 of reference [1].

7.3 Deep Space Optical Communication Simulation Results and Analysis

Simulations are done to investigate the impact of a varying PPM order (M = 2, 4, 8, 32, 64, 128, 256, 512, 1024, 2048 and 4096) on the number of received signal photons, noise photons, channel capacity, bit error probability, bit error rate and data rates in relation to the distance range dependency. All other parameters remained unchanged for the simulation. The figures below show the results of the simulations.

Figure 7.2 below shows that the number of detected signal photons decrease with increasing distance for a particular constant PPM order. Concurrently, the figure also depicts that for a specific constant distance range, higher PPM orders give rise to higher photon counts. Figure 7.3 below gives a 3D pictorial view of the interrelationship between channel capacity and BEP with varying distance range for different values of PPM order.

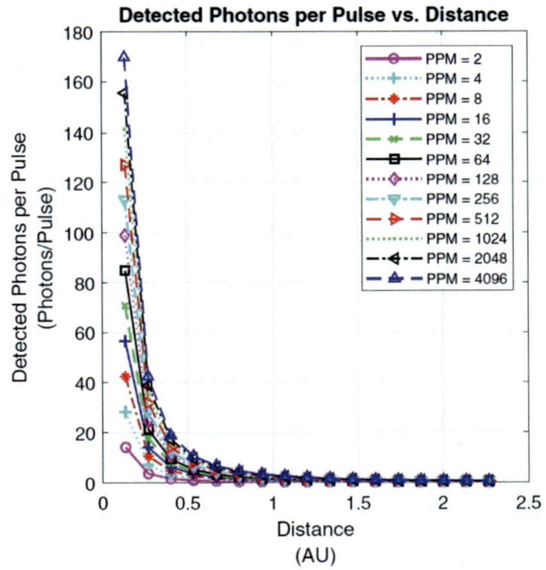

Fig. 7.2 Detected photons per pulse versus distance

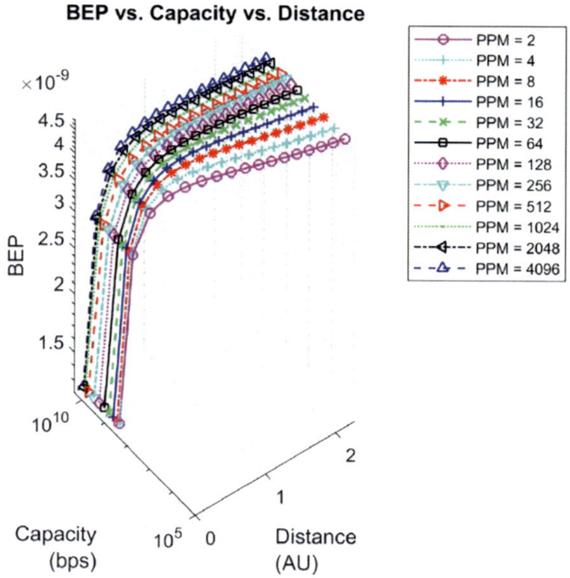

Fig. 7.3 Bit error probability (BEP) versus distance

Figures 7.4 and 7.5 indicate how the bit error rate (BER) varies with the number of photons/pulse and the number photons/slot respectively for different PPM orders. It can be observed that for a constant PPM order, BER decreases with increasing photons/pulse and photons/slot respectively. Moreover, as Hamid explained in page 254 of reference [1], with the presence of background noise, the dependence on M becomes evident as the BER performance for increasing M degrades.

Consequently, it is not appropriate to interpret BER performance with only photons/pulse as a measure of power efficiency as Fig. 7.4 shows. Thus, the use of the average number of photons/slot as a measure of power efficiency, as shown in Fig. 7.5, that takes into account the dependence of M when BER is plotted against photons (measure of power efficiency) gives an extra insight into the BER performance. Whereas low values of PPM order M in Fig. 7.4 produces a lower BER compared to high values of PPM order M, the contrasting scenario as shown in Fig. 7.5 occurs for different values of PPM order M. The complete analysis, that takes into account the dependence on M, when measuring BER performance versus average power efficiency in terms of photons as shown in Fig. 7.5, indicates that higher PPM orders M give better BER performance. Thus, taking a particular ratio photons/slot (a measure of average power efficiency) in Fig. 7.5 into consideration, higher PPM orders M provide lower BER. Similar explanation is attributed to charts from Figures 7.6 and 7.7 for BEP performances.

Figures 7.8 and 7.9 below respectively show the dependence of bit error rate (BER) and bit error probability (BEP) on distance for various values of PPM order M. It can be observed that for a specific constant PPM order M, BER and BEP increase with increasing distance.

Fig. 7.4 Bit error rates (BER) versus photons per pulse

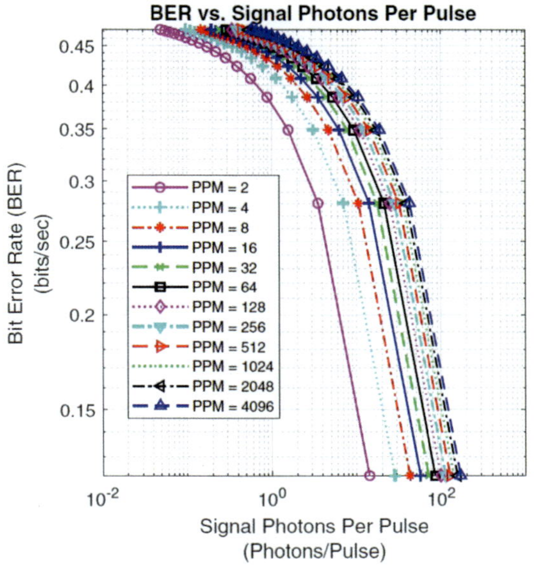

Fig. 7.5 Bit error rates (BER) versus photons per slot

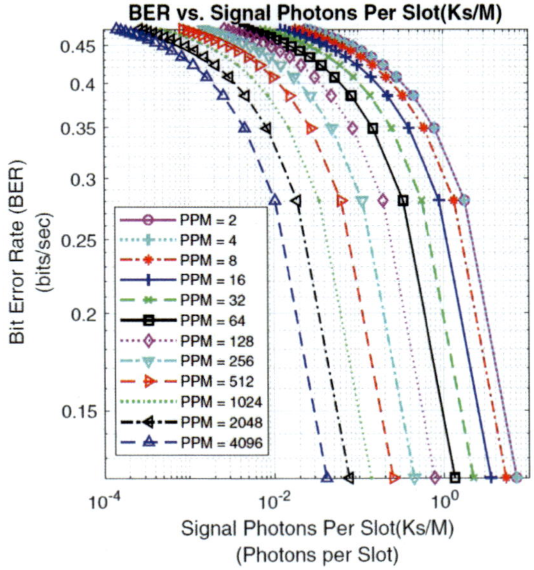

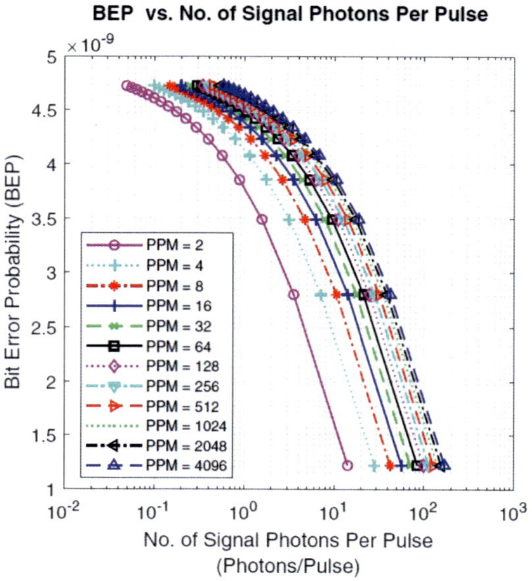

Fig. 7.6 Bit error probability (BEP) versus photons per pulse

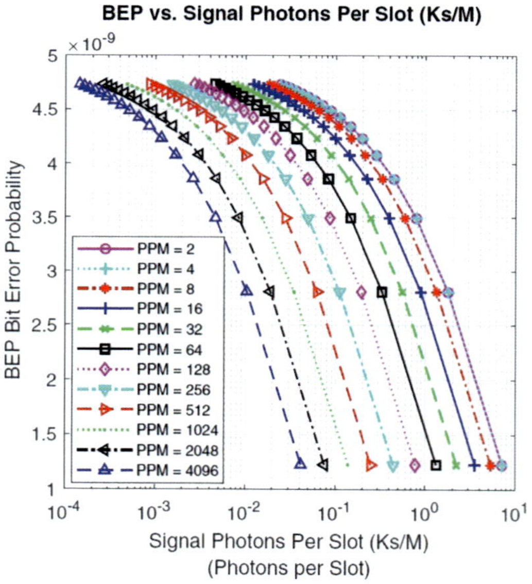

Fig. 7.7 Bit error probability (BEP) versus photons per slot

Moreover, it can be inferred that for a specific constant distance, higher values of the PPM order M provide lower BER and BEP in simulations of varying PPM order M as shown in Figs. 7.8 and 7.9 respectively. These charts and analyses are further explained in Figs. 7.5 and 7.7 where higher PPM orders M provided lower BER and BEP respectively.

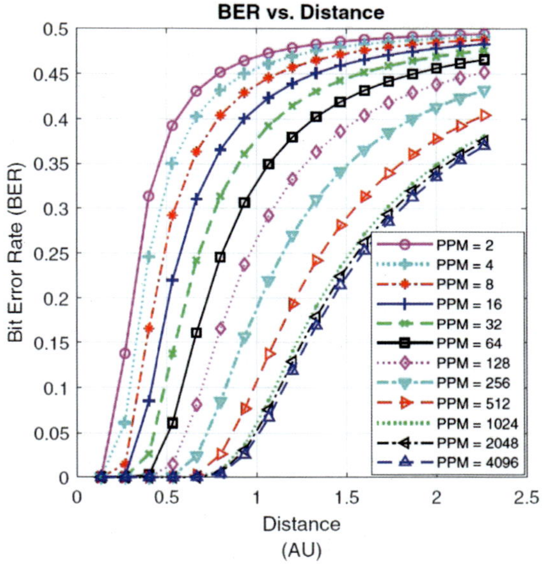

Fig. 7.8 Bit error rates (BER) versus distance

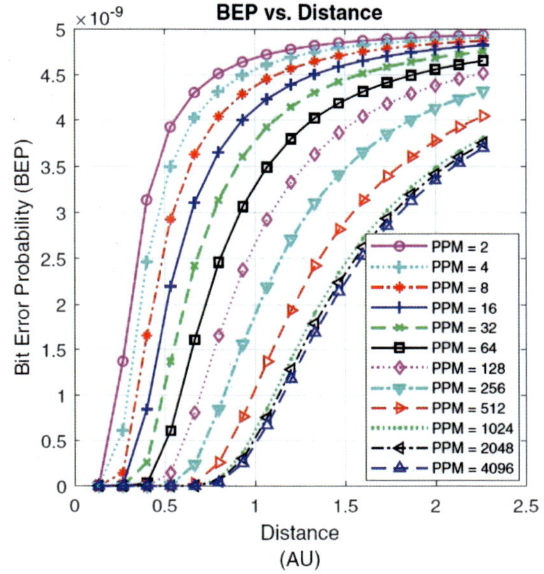

Fig. 7.9 Bit error probability (BEP) versus distance

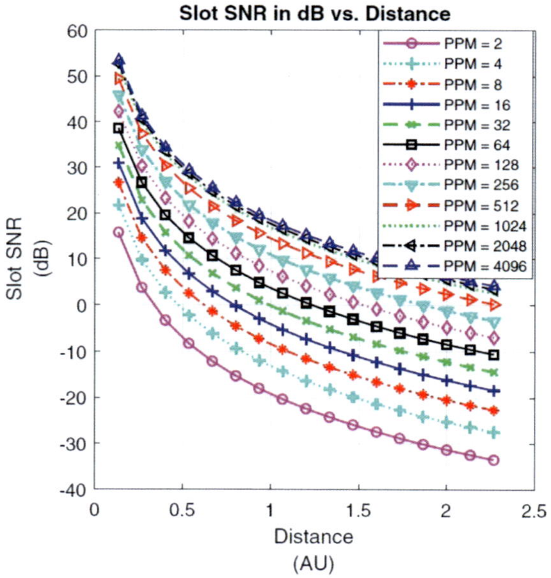

Fig. 7.10 Slot signal-to-noise ratio (SNR) versus distance

The dependence on distance is determined by plotting slot SNR and bit SNR against the distance in Figs. 7.10 and 7.11 for various PPM orders M respectively. Slot SNR is the signal to noise ratio for a symbol as a whole whereas bit SNR takes into account the number of bits in the symbol.

According to [1], slot SNR is calculated with the formula K_s^2/K_b whereas bit SNR is calculated using the formula $K_s^2/(2K_b \cdot log_2 M)$ for any given set of values of signal photons (K_s), background photons (K_b), and PPM order (M).

Consequently, slot SNR shows less dependence on the PPM order M. Furthermore, the charts in Figs. 7.10 and 7.11 show that for a specific constant PPM order M, both slot SNR and bit SNR decrease with increasing distance. Moreover, it can be inferred from the results in Figs. 7.10 and 7.11 below that considering a specific constant distance, there exist an average of a 30 dB gap between SNR for simulations when varying the PPM order from M = 2 to M = 4096. Thus, higher PPM orders have a higher slot SNR and bit SNR for a specific constant distance.

The PPM symbol error probability (SEP) versus photon/slot is shown below in Fig. 7.12 whereas Fig. 7.13 exhibits the PPM symbol error probability (SEP) dependency on the distance. According to references [1, 5, 12], the PPM symbol error probability (SEP) relates to bit error probability by the equation given in page 254 of [1] as

$$P_b = 0.5 \cdot P_s \cdot \frac{M}{(M-1)} \tag{7.22}$$

Figure 7.12 indicates that for a specific constant PPM order (M), SEP decreases with increasing photons/slot. Also, considering a specific constant photons/slot in Fig. 7.12, it

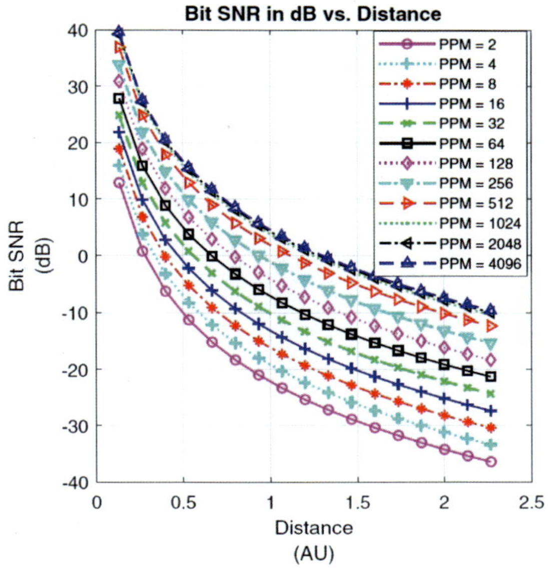

Fig. 7.11 Bit signal-to-noise ratio (SNR) versus distance

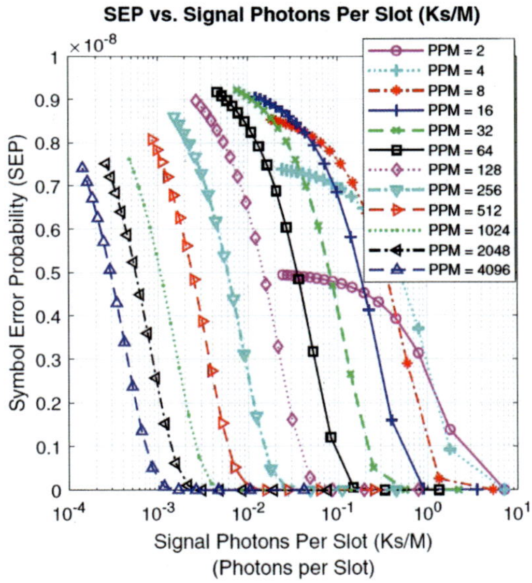

Fig. 7.12 Symbol error probability (SEP) versus photons per slot

is observed that the SEP decreases with increasing PPM order from M = 8 to M = 4096. Moreover, it can be inferred from Fig. 7.13 that SEP increases with increasing distance. Also, considering a specific constant distance in Fig. 7.13, it can be seen that the SEP decreases with increasing PPM order from M = 8 to M = 4096.

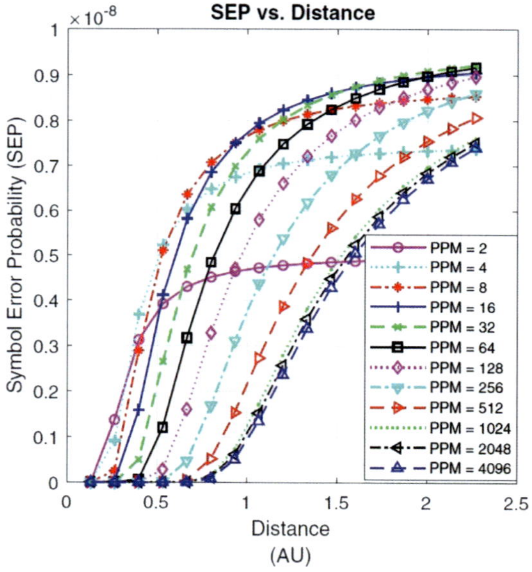

Fig. 7.13 Symbol error probability (SEP) versus distance

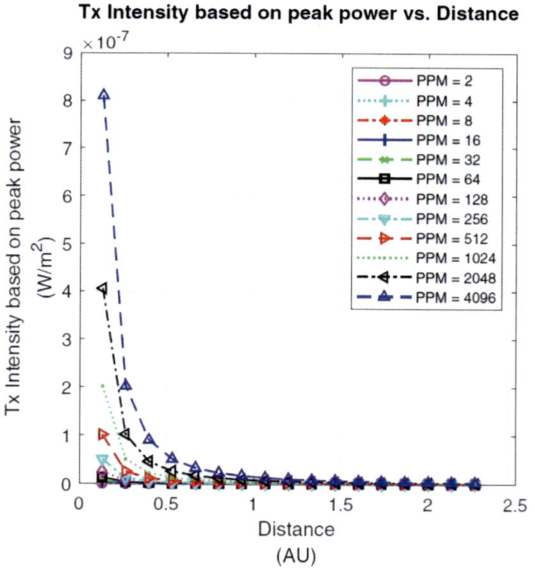

Fig. 7.14 Intensity versus distance

Figure 7.14 below shows that the intensity (based on the peak power per symbol) varies according to the PPM order whereas Fig. 7.15 shows that the effective area of divergence is unaffected by the variation of the PPM order. Concurrently, it can be seen that the transmit intensity of the laser reduces with increasing distance whereas the effective area of divergence increases with increasing distance.

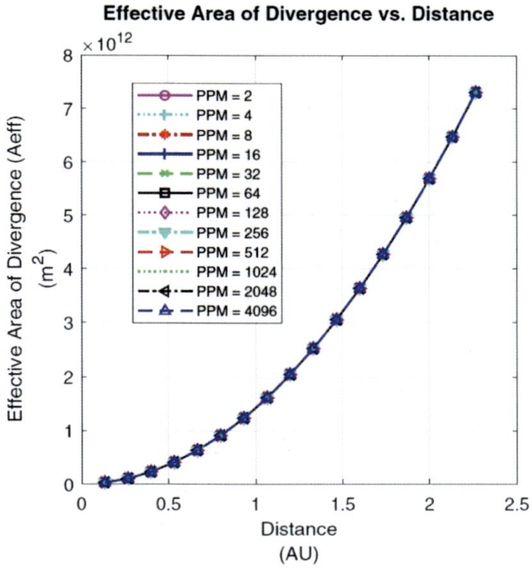

Fig. 7.15 Beam divergence versus distance

The capacity of the optical channel is a very important reference value for determining the useful data rates achievable with any modulation and coding scheme. Thus, the data rate can be determined directly from the channel capacity taking into consideration the code rate factor; hence the need to determine how the optical channel capacity changes with varying PPM order (M). According to [1], the capacity of the optical channel is a function of the received optical signal and noise powers, the modulation and the detection method. A hard-decision capacity is considered whereby the receiver makes estimates of each PPM symbol, passing these estimates, or hard decisions, on to the decoder. Consequently, the (hard-decision) capacity can be expressed as well as a function of the probability of symbol error Ps.

Figure 7.16 shows that the optical channel capacity increases with increasing average received power. Considering the chart produced by each PPM order, Fig. 7.16 also exhibits three segments for the plot of channel capacity versus average received power, that is, noise-limited capacity (in the form of quadratic plot in the beginning of the chart), quantum-limited capacity (in the form of linear plot in the middle of the chart) and the bandwidth-limited capacity (in the form of a saturated or constant plot at the tail-end of the chart) following in this order along an increasing trend of the average received power. According to [1, 13], there is a peak constraint in the power resulting in an upper limit on the PPM order and hence the upper limit (saturation) in optical channel capacity.

Figures 7.17, 7.18 and 7.19 respectively show how the optical channel capacity, data rate and data throughput change with distance for various values of the PPM order M. Generally, for a specific constant PPM order M, the optical channel capacity, data rate and data throughput decrease with increasing distance as shown in Figures. Moreover, as shown

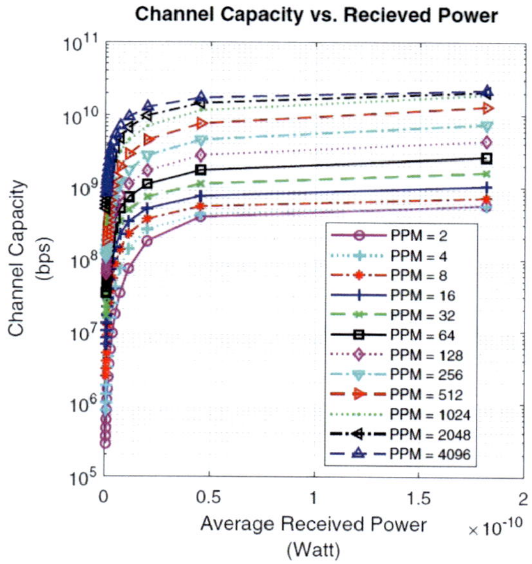

Fig. 7.16 Channel capacity versus received power

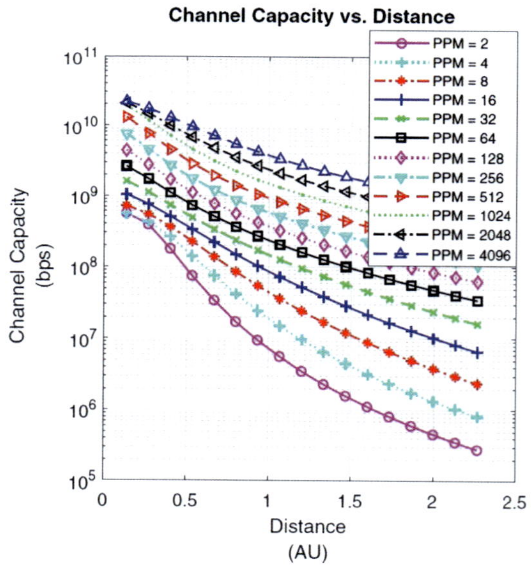

Fig. 7.17 Channel capacity versus distance

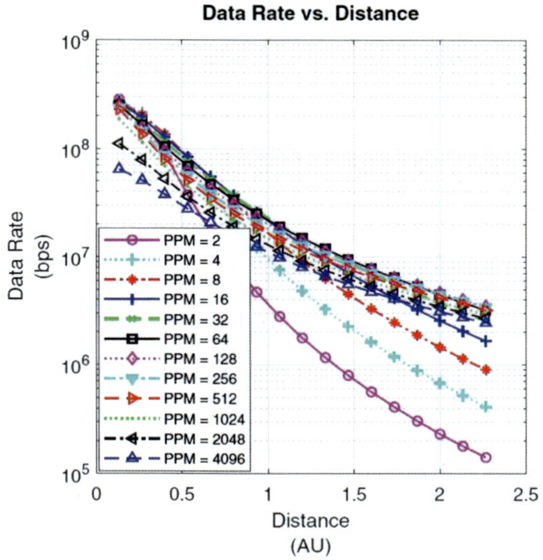

Fig. 7.18 Data rate versus distance

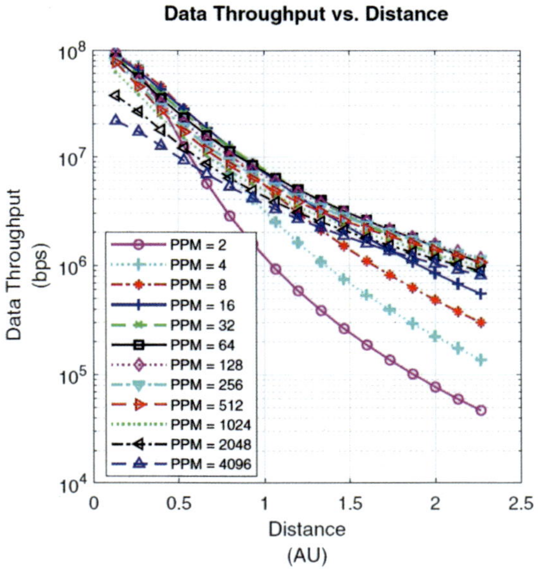

Fig. 7.19 Data throughput versus distance

in the charts for the data rate and data throughput for distance ranges less than 0.7AU, PPM orders M less than 2048 produce fairly better performance in throughput. This is explained by the peak capacity constraint (saturation) that occurs due to an upper limit on the PPM order. Furthermore, for distance ranges more than 0.7 AU, PPM orders from M = 2 to M = 16 produce degraded performances in data rate. Generally, it can be observed that above 1 AU, as depicted in Fig. 7.18 because increasing the PPM order from M = 32 to M = 4096 does not produce a sharp increment in the data rate compared to increasing the PPM order from M = 2 to M = 32. This is the saturation limit on the data rate with respect to the upper limit constraint in PPM order M as explained earlier. This also partly explains why very high PPM orders like M = 2048 or M = 4096 are not practically used in Earth-Mars missions. These very high PPM orders also come with other requirements such as very high computing processing resources and peak power.

As stated in the introduction, an Earth-Mars mission lifetime is associated with varying distances over an orbital period ranging from 0.37 AU (56 million km) to 1.5 AU (225 million km); hence a robust PPM order M must be chosen to take into account the performance of the data rate within this distance ranges and conditions. Consequently, all these deductions from Figs. 7.18 and 7.19 help to understand reasons why PPM orders of 16/64/128/256 are commonly used by many deep space missions among the range of 2-4096 in order to achieve an optimum high data rate. Although up to M = 256 is used till now, due to different technical reasons (like compute processing resources and peak power), the possibility of using M up to 1024 in the future could show some further improvement, as can be seen in the simulation results.

7.4 Conclusions

Simulation results have confirmed that optimization to improve the received signal power (or received signal photons) can be accomplished by varying the PPM order M. This can help to improve the ratio of the received signal photons to noise photons. Thus, increasing M can improve the capacity, bit error probability, bit error rate and achievable data rate of a DSOC system. Nevertheless, that leads to less bandwidth efficiency. In addition to the effective delivery of the signal to the detector, the performance of the optical link also depends on the receiver sensitivity (measured in terms of received photons per bit). Because of high cost associated with increasing transmit power and system aperture, improving the receiver sensitivity is an important factor in deep space optical communication system design. Thus optimizing the value of the PPM order (by choosing high M) could be the best way of achieving improved capacity, data rates and bit error probability. Nevertheless, the dead-time of the APD single photon counting receiver will affect the total slot width of the PPM codes, and should be considered in the design. Also the very high PPM order comes with requirements of high computing processing resources and also a strict slot synchronization and error correction to ensure reliable data transmission and recovery.

Finally, simulation results have helped to explain reasons why PPM orders of 16/64/128/256 are commonly used by many deep space missions among the range of 2-4096 in order to achieve high data rates and optimum channel capacity. Even though M up to 256 is used till now, due to different technical reasons (like compute processing resources and peak power), the possibility of using M up to 1024 for deep space in the future could show some further improvement, as can be seen in the simulation results.

References

1. H. Hemmati, "Deep Space Optical Communications", Deep Space Communications and Navigation Systems, Center of Excellence, Jet Propulsion Laboratory, California Institute of Technology, Publisher: Wiley 2006.
2. A. Biswas, K. E. Wilson, S. Piazzolla, J. Wu, W. H. Farr, "Deep Space Optical Communications Link Availability and Data Volume", Deep Space Communications and Navigation Systems, Center of Excellence, Jet Propulsion Laboratory, California Institute of Technology, The International Society for Optical Engineering, pp. 1–10, June 2004.
3. A. Biswas, and S. Piazzolla, "Deep-Space Optical Communications Downlink Budget from Mars: System Parameters", The Interplanetary Network Progress Report, pp. 1–38, June 2003.
4. R. Gallager, "Information Theory and Reliable Communication", New York: Wiley, 1968.
5. J. Hamkins, "Accurate Computation of the Performance of M-ary Orthogonal Signaling on a Discrete Memoryless Channel", IEEE Transactions on Communications, 2003.
6. C. W. Helstrom, "Quantum Detection and Estimation Theory", Mathematics in Science and Engineering, vol. 123, New York, Academic Press, 1976.
7. E. D. Aboagye, S.-P. Chen: Deep space optical communications (DSOC) downlink simulation with varying PPM order. Optical and Quantum Electronics, 53, 583 (2021). https://doi.org/10.1007/s11082-021-03232-z
8. C.-C. Chen, "Figure of Merit for Direct-Detection Optical Channels", The Telecommunications and Data Acquisition Progress Report 42-109, January - March 1992, Jet Propulsion Laboratory, Pasadena, California, pp. 136–151, May 15, 1992.
9. S. Arnon, S. R. Rotman, and N. S. Kopeika, "Performance limitations of free-space optical communication satellite networks due to vibrations: direct detection digital mode," OE - Optical Engineering, vol. 36, no. 11, pp. 3148 - 3157, Nov 1997.
10. A. D. Wyner, "Capacity and Error Exponent for the Direct Detection Photon Channel - Part I", IEEE Transactions on Information Theory, vol. 34, no. 6, pp. 1449–1961, November 1988.
11. J. R. Pierce, E. C. Posner, and E. R. Rodemich, "The Capacity of the Photon Counting channel", IEEE Transactions on Information Theory, vol. IT-27, no. 1, pp. 61–77, January 1981
12. L. W. Hughes, "A Simple Upper Bound on the Error Probability for Orthogonal Signals in White Noise", IEEE Transactions on Communications, vol. 40, p. 670, April 1992
13. B. Moision and J. Hamkins, "Deep-Space Optical Communications Downlink Budget: Modulation and Coding", The Interplanetary Network Progress Report 42-154, April-June 2003, Jet Propulsion Laboratory, Pasadena, California, pp. 1-28, August 15, 2003.

8 Lunar and Deep Space Optical Communications

The link performance of free space and deep space optical communications (FSOC and DSOC) was investigated by using algorithms coded with MATLAB and by considering the key mission design parameters such as laser transmitter power, operating wavelength, telescope aperture sizes, distance between ground station and spacecraft, modulation scheme and background noise [1–5]. With recently published measurement data from the on-going NASA/ESA's Psyche mission [6, 7], the previously published data from NASA's "Lunar Laser Communication Demonstration (LLCD)" project [8, 9], and other DSOC system parameters in [10], the simulation algorithms can be validated for both relatively short distance and large distance.

8.1 Introduction

Typical laser optical communication links are discussed ranging from near-Earth, lunar to interplanetary laser optical communications. The performance for different scenarios has been investigated by using simulation models which were successfully verified in previous projects. The simulation models consider typical mission design parameters like transmit power and operating wavelength for the optical laser transmitters, different apertures of the optical telescope, varying distances between the Earth ground station and the satellite or spacecraft terminals. In case of deep space optical communications, because of the extremely large distance and the extremely large attenuation leading to few numbers of received photons, photon counting receivers are indispensable. Atmospheric disturbances like scintillation, absorption, and background noises for uplink and downlink are considered. For a relatively short distance, higher order modulation schemes like M-QAM can be applied and higher bitrates can be achieved, whereas in higher orbits or at an extremely large

S.-P. Chen, *Fundamentals of Free Space and Deep Space Optical Communications*,
https://doi.org/10.1007/978-3-032-35826-4_8

distance, appropriate pulse position modulation (PPM) orders M will be necessary. Typical design parameters, especially for higher orbits or at large distance, are those of published ESA (European Space Agency) and NASA (National Aeronautics and Space Administration) missions for GEO-orbits, lunar and interplanetary orbits. The same method can be also used to investigate optical Inter-Satellite-Links.

8.2 Lunar Laser Optical Communications

In Lunar Optical Communications PPM modulation will be utilized with M = 4 for uplink and M = 16 for downlink (details in [2–5] and in Tables 8.1, 8.2.

Table 8.1 Simulation parameters for lunar laser communications, uplink [8, 9]

Lunar uplink	350,000–410,000 km $2.33 \cdot 10^{-3}$ AU - $2.73 \cdot 10^{-3}$ AU
Pulse position modulation PPM order M	4
Code rate CR	1/2
TX aperture	4 x 0.4 m
TX power	4 x 10 W
TX laser diode wavelength	1550 nm
TX efficiency	0.7
Atmospheric transmission	0.8
RX aperture efficiency with filter	0.16
RX aperture diameter	0.1 m
Required irradiance at spacecraft, 5 nm filter	63 nW/m^2 or 1.27 nW

Table 8.2 Simulation parameters for lunar laser communications, downlink [8, 9]

Lunar downlink	350,000–410,000 km $2.33 \cdot 10^{-3}$ AU - $2.73 \cdot 10^{-3}$ AU
Pulse position modulation PPM order M	16
Code rate CR	1/2
TX aperture	0.1 m
TX power	500 mW
TX laser diode wavelength	1558 nm
TX efficiency	0.7
Atmospheric transmission	0.73
RX aperture efficiency with filter	0.16
RX aperture diameter	4 × 0.4 m
Required irradiance at ground station	1.7 nW/m^2 or 0.21 nW
Blue sky brightness, daytime	41.2 pW/m^2 or 32.4 pW
Lunar background	4.2–132.1 pW/m^2 or 0.53–16.6 pW

Tables 8.1 and 8.2 show the lunar laser optical communications simulation parameters for uplink and downlink. Basically the parameters are similar to the NASA LLCD project in cooperation with ESA [8, 9]. In Fig. 8.1, Figs. 8.2 and 8.3, the uplink characteristics, like power intensity of the laser beam at the receiver, received signal photons and achievable data throughput are displayed, whereas Figs. 8.4, 8.5 and 8.6 show the downlink characteristics. As an example the PPM orders 4 and 16 with a code rate of 1/2 are used to estimate the throughput for uplink and downlink communications.

Fig. 8.1 shows that uplink power intensity at the receiver onboard the spacecraft, with additional deviation $\pm\sigma_I$ caused by the atmospheric scintillation in the troposphere up to about 11 km, decreases to about 0.35 μW/m^2 for a larger distance up to 0.003 AU. But it is strong enough compared with the required irradiance at the spacecraft of 63 nW/m^2. The average numbers of the signal photons per pulse are about 99 (Fig. 8.2). A maximum bitrate of about 15 Mbit/s can be achieved correspondingly by using the link parameters from Table 8.1 (Fig. 8.3).

Fig. 8.4 shows that downlink power intensity at the receiver at the ground station, with additional deviation $\pm\sigma_I$ caused by the atmospheric scintillation in the troposphere up to about 11 km, decreases to about 7.8 nW/m^2 for a larger distance up to 0.003 AU. But it is still strong enough compared with the required irradiance at the ground station of 1.7 nW/m^2. The average numbers of the signal photons per pulse are about 99 (Fig. 8.5). A bitrate of about 466 Mbit/s can be achieved correspondingly by using the link parameters from Table 8.2 (Fig. 8.6).

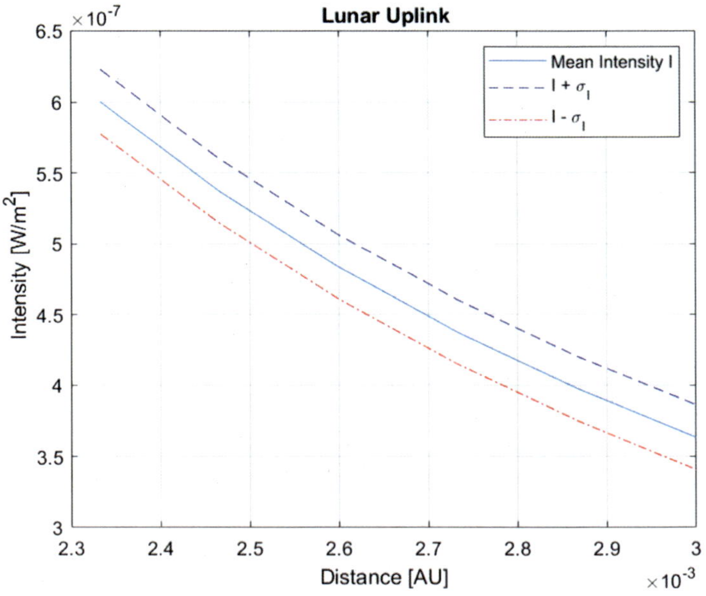

Fig. 8.1 Lunar uplink intensity at the receiver versus distance based on the parameters in Table 8.1

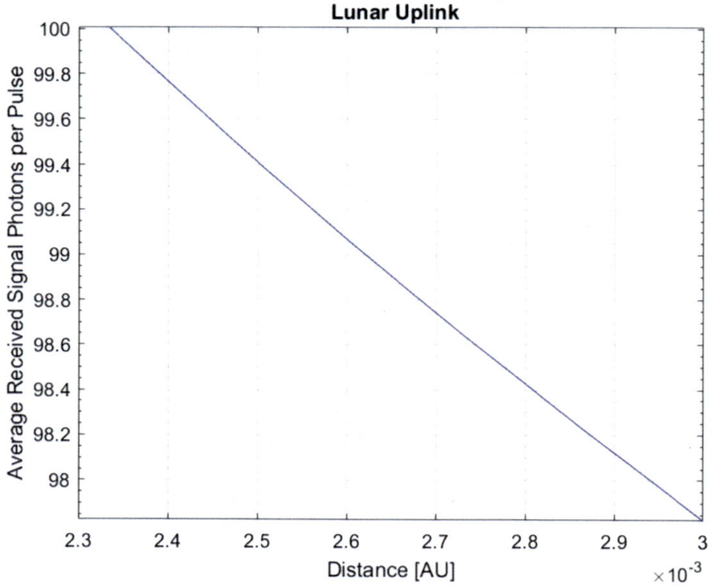

Fig. 8.2 Lunar uplink received signal photons per PPM pulse versus distance based on the parameters in Table 8.1

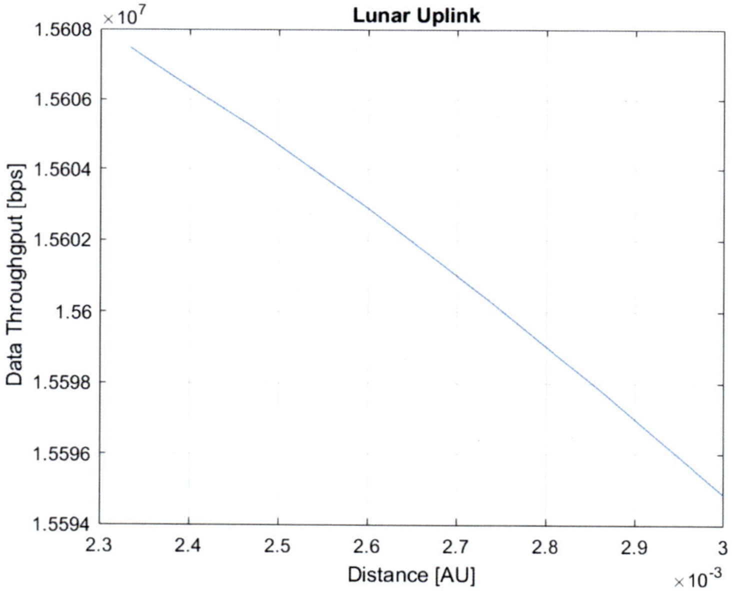

Fig. 8.3 Lunar uplink maximum data rate versus distance based on the parameters in Table 8.1

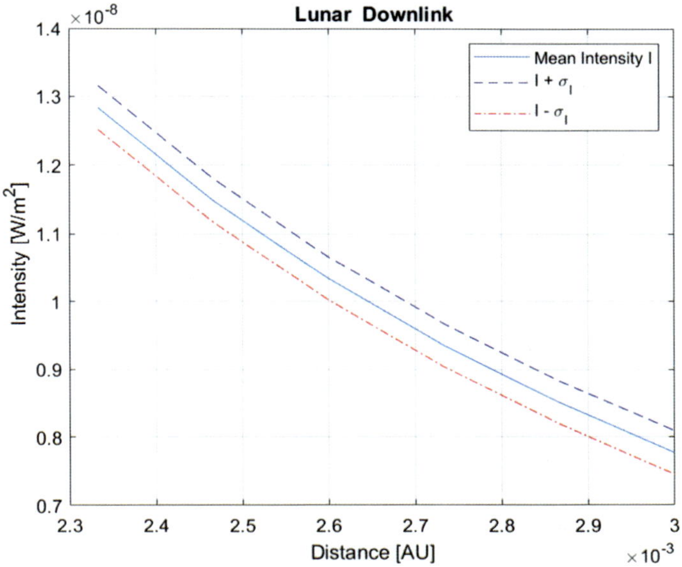

Fig. 8.4 Lunar downlink intensity at the receiver versus distance based on the parameters in Table 8.2

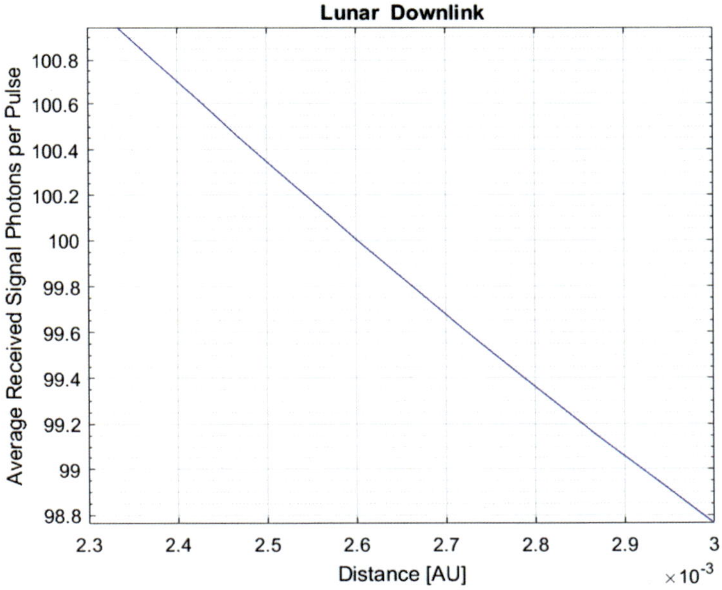

Fig. 8.5 Lunar downlink received signal photons per PPM pulse versus distance based on the parameters in Table 8.2

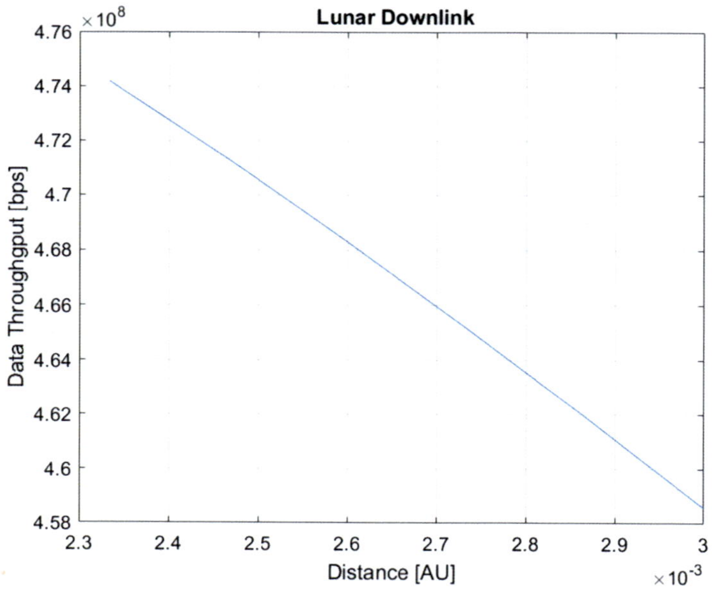

Fig. 8.6 Lunar downlink maximum data rate versus distance based on the parameters in Table 8.2

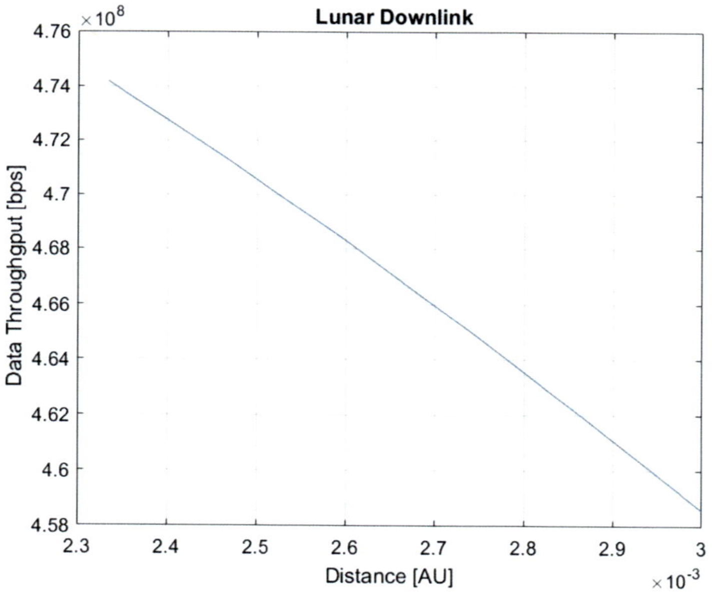

Fig. 8.7 Lunar laser optical communications downlink. Data throughput versus distance

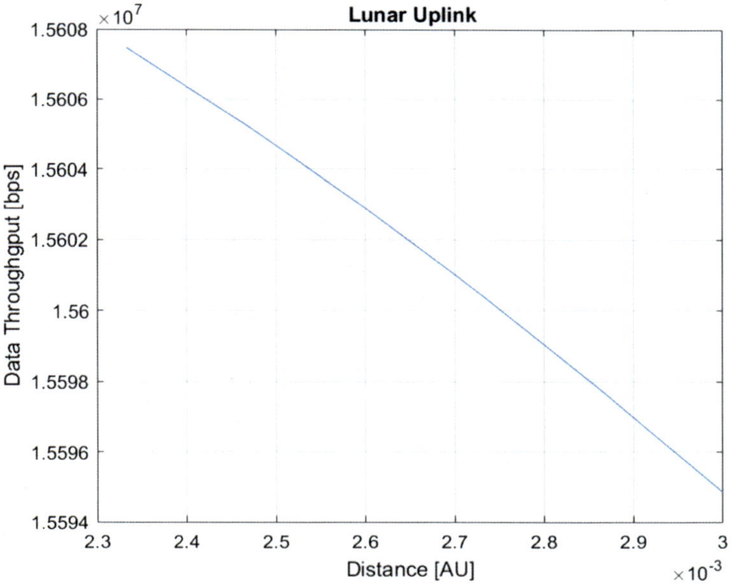

Fig. 8.8 Lunar laser optical communications uplink. Data throughput versus distance

8.3 Deep Space Optical Communications and Interplanetary Laser Optical Communications

Due to the extremely long distance and strong attenuation, the signal-noise-ratio for DSOC (Deep Space Optical Communications) becomes very poor, so that higher modulation schemes cannot be applied. A very robust and reliable modulation scheme is PPM (Pulse Position Modulation) which is utilized with M = 2–4096 (Figs. 8.7 and 8.8).

The simulation parameters for interplanetary or deep space laser optical communications are shown in Table 8.3 and Table 8.4. These are basically derived from the ESA DOCS L5 parameters [7]. In Figs. 8.9, 8.10 and 8.11 the uplink characteristics, like power intensity of the laser beam and achievable data throughput are displayed, whereas Figs. 8.12, 8.13 and 8.14 show the downlink characteristics. As example the PPM orders 4 and 16 with a code rate of 1/2 are used to estimate the throughput for uplink and downlink communications.

Especially in the photon-starved laser optical transmission over an extremely large distance of 1.5 AU to 3 AU, it is most important to apply high laser power and larger receiver telescope apertures to focus the beam at the transmitter with high gain, and to collect as many photons as possible at the receiver terminals, in order to guarantee sufficient signal-to-noise ratio and to enable the photon counting device to achieve low BER and thus high throughput. Transmitters can use a coherent beam combiner with 4 apertures of 0.2 m, or 8

Table 8.3 Simulation parameters for interplanetary laser communications, uplink [7, 10]

Interplanetary uplink	0.5–3 AU
Pulse position modulation PPM order M	4
Code rate	1/2
TX aperture	8 × 0.1 m
TX power	2.5 kW
TX laser diode wavelength	1064 μm
TX efficiency	0.7
Atmospheric transmission	0.8
RX aperture efficiency with filter	0.7
RX aperture diameter	0.22 m
Earth background brightness, 5 nm filter at 1.5 AU	36.4 pW/m^2 or 0.8 pW

Table 8.4 Simulation parameters for interplanetary laser communications, downlink [7, 10]

Interplanetary downlink	0.5–3 AU
Pulse position modulation PPM order M	16
Code rate	1/2
TX aperture	0.22 m
TX power	4 W
TX laser diode wavelength	1550 μm
TX efficiency	0.7
Atmospheric transmission	0.73
RX aperture efficiency with filter	0.5
RX aperture diameter	4 m / 10 m
Blue sky brightness, 3 nm filter	0.08 pW/m^2 or 0.5 pW

apertures of 0.1 m diameter, or a larger transmitter telescope apertures in combination with adaptive optics, where applicable.

Figure 8.9 shows that uplink power intensity at the receiver onboard the spacecraft, with additional deviation $\pm\sigma_I$ caused by the atmospheric scintillation in the troposphere up to about 11 km, decreases from 3 nW/m^2 to about 0.1 nW/m^2 for a larger distance up to 3 AU. Even though the mean intensity is enough compared with the Earth background irradiance of 36.4 pW/m^2 at 1.5 AU, the atmospheric scintillation could reduce the irradiance significantly for larger distance. The average numbers of the signal photons per pulse are between 84 and 70 (Fig. 8.10). A maximum bitrate of some 14 Mbit/s at 0.5 AU to 212 kbit/s at 3 AU can be achieved correspondingly by using the link parameters from Table 8.5 (Fig. 8.11).

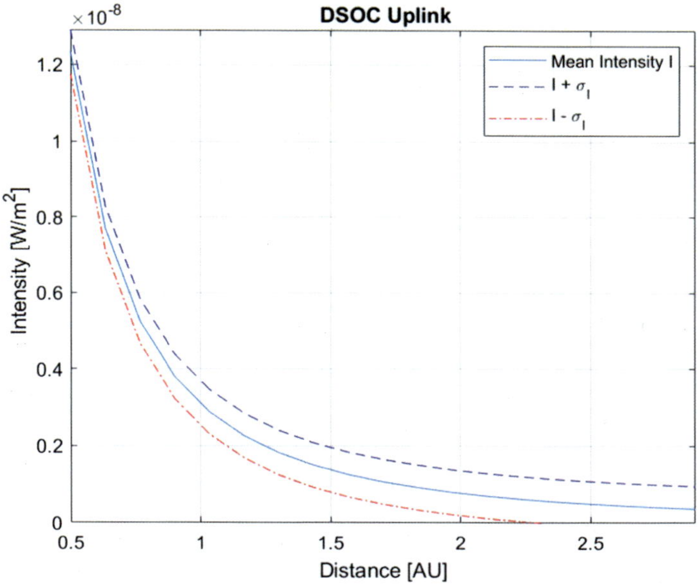

Fig. 8.9 DSOC uplink intensity at the receiver versus distance based on Table 8.3

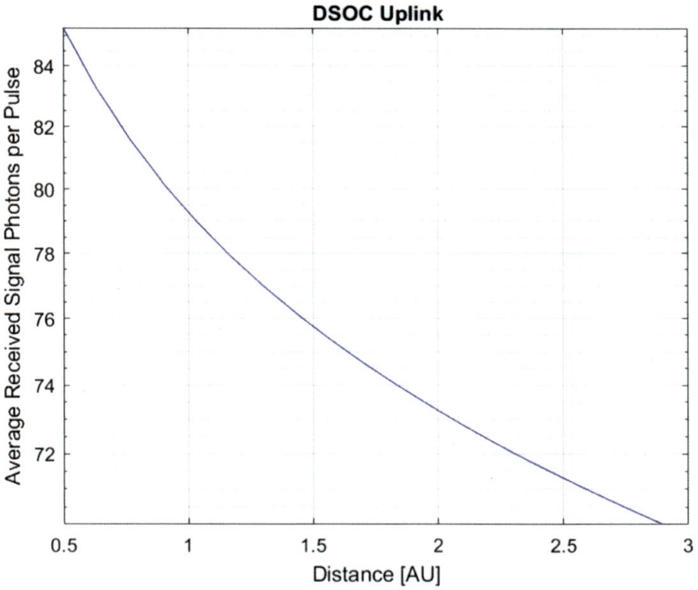

Fig. 8.10 DSOC uplink received signal photons per PPM pulse versus distance based on Table 8.3

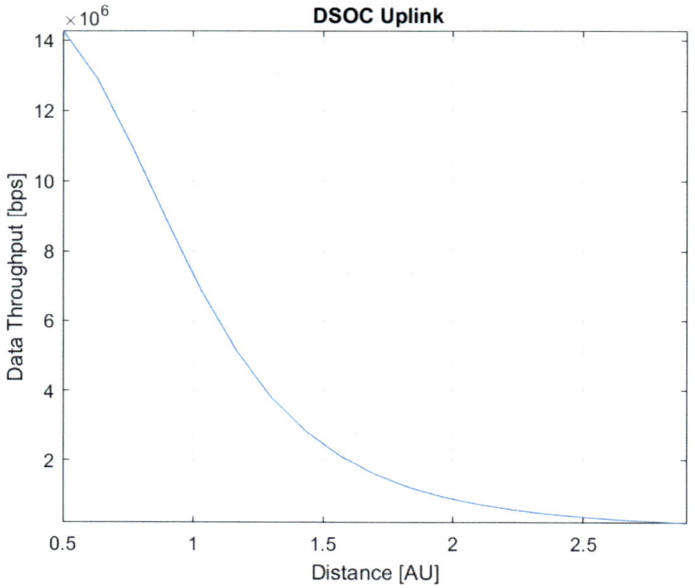

Fig. 8.11 DSOC uplink maximum data rate versus distance based on Table 8.3

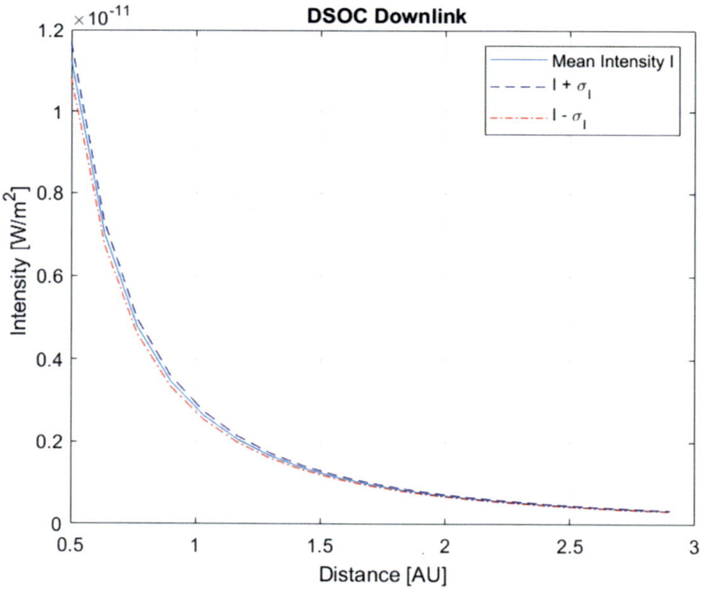

Fig. 8.12 DSOC downlink intensity at receiver versus distance based on Table 8.4

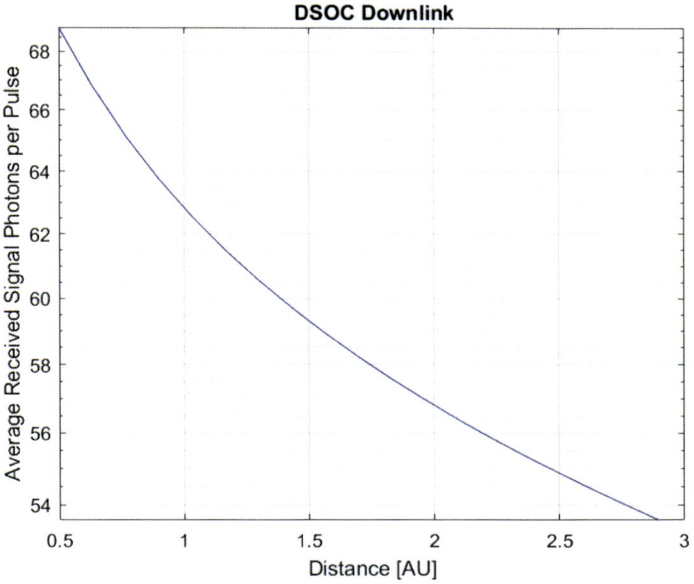

Fig. 8.13 DSOC downlink received signal photons per PPM pulse versus distance based on Table 8.4

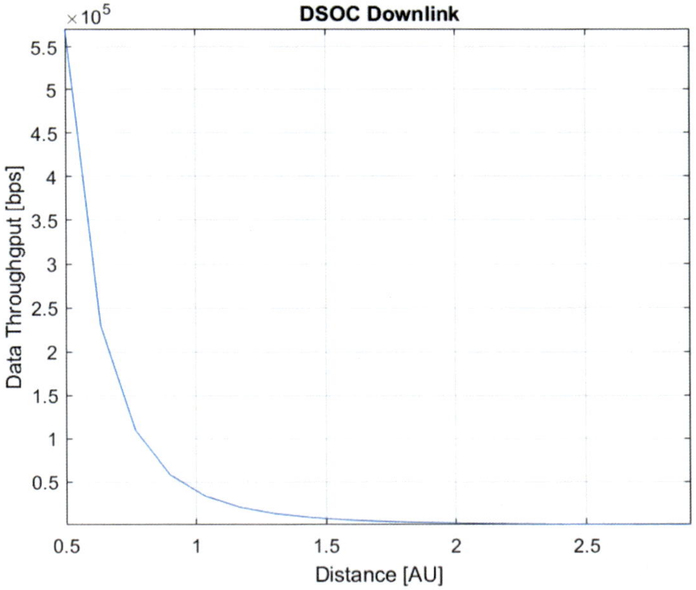

Fig. 8.14 DSOC downlink maximum data rate versus distance based on Table 8.4 with 4 m ground station receiver aperture

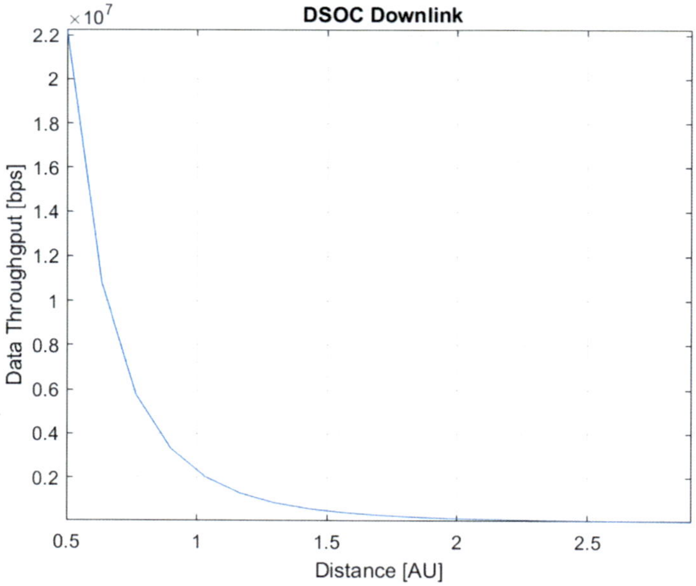

Fig. 8.15 DSOC downlink data throughput versus distance based on Table 8.4 with 10 m ground station receiver aperture

Figure 8.12 shows that downlink power intensity at the receiver at the ground station, with additional deviation $\pm\sigma_I$ caused by the atmospheric scintillation in the troposphere up to about 11 km, decreases from 11 pW/m^2 to about 0.3 pW/m^2 for a larger distance up to 3 AU. Even though the mean intensity is still enough compared with the blue-sky irradiance at the spacecraft of 0.08 pW/m^2, the atmospheric scintillation could reduce the irradiance significantly for distance larger than 2 AU. The average numbers of photons per pulse are between 68–54 (Fig. 8.13). A maximum bitrate of about 0.57 Mbit/s at 0.5 AU to 559 bit/s at 3 AU can be achieved correspondingly by using a 4 m ground station receiver aperture from Table 6 (Fig. 8.14). If however a larger telescope aperture of 10 m could be used, then the maximum bitrates can be improved to 22 Mbit/s at 0.5 AU to 39 kbit/s at 3 AU (Fig. 8.15).

Figure 8.16 shows the channel capacity versus distance for different PPM order M, whereas Fig. 8.17 shows the achievable data throughput versus distance depending on the PPM order M (Fig. 8.17).

8.4 Links Between Lagrange L5 and Earth Ground Station

ESA plans to embark on the Space WEather (SWE) L5 mission to the Sun-Earth Lagrange point L5 in the frame of its Space Situational Awareness (SSA) program [7]. This mission will use Deep Space Optical Communications System (DOCS), an in-orbit technology

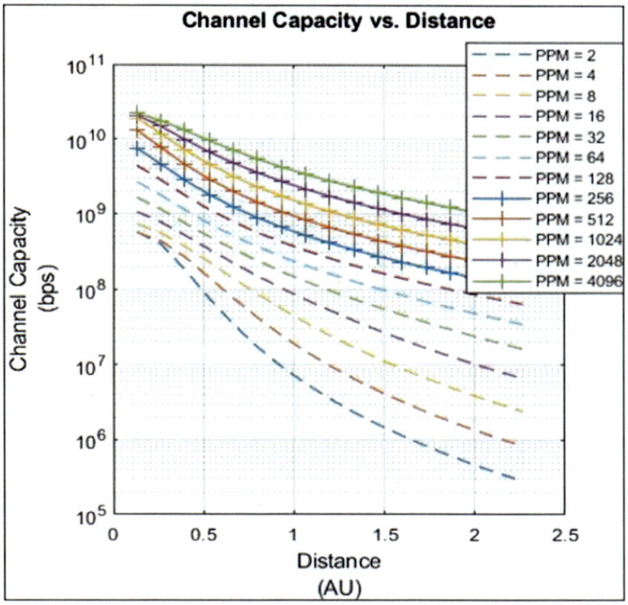

Fig. 8.16 DSOC channel capacity for different PPM order versus distance

Fig. 8.17 DSOC throughput for different PPM order versus distance

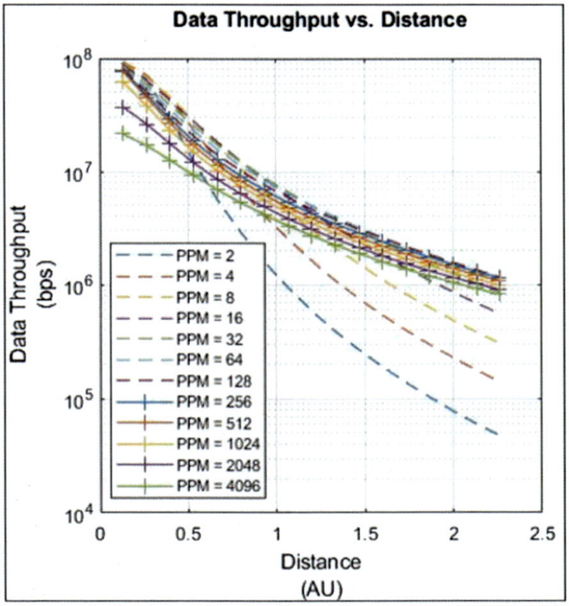

demonstration that also serves a scientific objective, namely the transfer of high-resolution solar imagery. The characteristics of the SWE L5 mission provide substantial advantages for deep space optical communications, because the equidistant triangular orbital geometry between Sun, Earth, and L5 (all three distances are equal to 1 AU $\approx$ 150 million km) ensures that the sun is always separated by 60 degrees from both the space and the ground terminal. This allows for very efficient solar stray-light shielding and thermal management. It also reduces the pointing requirements of the DOCS Space Terminal (DST). A coarse pointing mechanism is not required, nor is a point-ahead assembly. Also in this project, the OGS in Tenerife, Spain will play an important role. It is called DOCS Ground Terminal (DGT).

The system parameters for downlink and uplink are listed in Tables 8.5 and 8.6. The comparison with the designed link budget of [7] is shown in Table 8.7. The independent simulation results in this work match with the link budget calculation pretty well. In DSOC, besides the atmospheric scintillation also other effects, like solar plasma or sun irradiance, would affect the laser transmission. Once the first DSOC measurement data are available, for example from the Psyche mission for distances up to 3 AU, it will be illuminating to

Table 8.5 Downlink DOCS DST to DGT

DOCS DST–DGT	1 AU $\approx$ 150,000,000 km
TX power	15 mW
TX laser diode wavelength	1550 nm
TX efficiency	0.7
TX aperture	20 cm
Atmospheric transmission	0.73
RX aperture efficiency with filter	0.5
RX aperture diameter	1 m or 4 m
Blue sky brightness, 3 nm filter	0.5 pW

Table 8.6 Uplink DOCS DGT to DST

DOCS DGT - DST	1 AU $\approx$ 150,000,000 km
TX power 4 $\times$	600 W
TX laser diode wavelength	1064 nm
TX efficiency	0.7
TX aperture 4 $\times$	10 cm
Atmospheric transmission	0.8
RX aperture efficiency with 5 nm filter	0.7
RX aperture diameter	20 cm
Earth background, 5 nm filter	0.8 pW

Table 8.7 Comparison of the designed link budget [7] and simulation in this work

	Downlink 4 m receiver	Downlink 1 m receiver	Uplink
DOCS design	9.30 pW	0.60 pW	8.50 pW
This work	9.43 pW	0.59 pW	9.02 pW

investigate the scintillation effects to further improve our simulation methods. Based on the results achieved up to now, we are optimistic that the simulation methods presented here can be further developed to more completely characterize optical communications for deep space missions at interplanetary distances.

8.5 Further Considerations and Steps for the Performance Analysis and Planning of Deep Space Optical Communications

As discussed above, it would be meaningful to collect laser link performance measurement data along with the critical weather data like wind velocity, temperature and pressure in the atmosphere in future mission experiments, in order to better understand the temporal changes of the mean irradiance and the scintillation effects.

For interplanetary deep space missions with distances of 1–3 AU, when the photons are extremely starved, it is of great importance to consider the penalty of the daytime measurement versus nighttime measurement, since the blue-sky background and Earth background noise could significantly affect the measurements. These effects must be considered and utilized in the future DSOC investigations correspondingly.

Another huge factor for mission planning is also the so-called SEP (Sun-Earth Probe angle), i.e. the closeness of the spacecraft to the Sun, as seen in the sky. Just like the radio frequency propagation will be disturbed by the solar plasma, the optical wave propagation will also be affected by the infrared radiation of the Sun significantly, leading to performance degradation. This effect must also be taken into account in future deep space mission planning.

For interplanetary DSOC with starved photons at the receivers, single photon counting APD is indispensable. Proper PPM orders have to be chosen, and the deadtime of the single photon counting APD must be taken into account.

The site dependencies are very influential to mission planners for designing DSOC systems. The weather conditions like wind velocity, clouds, and elevation angles are crucial factors. The above-mentioned simulation method for the link budget analysis and the scintillation estimation will be combined, further improved in a proper way for the analysis of future missions.

As documented in this book, we have verified our simulation methods by comparison with existing mission planning design data DOCS L5 and ARTEMIS OGS OPALE measurement results, under somewhat constrained or limited conditions. With these successful verifications, we plan to further improve the simulation algorithms to more completely represent additional factors that impact the performance of deep space optical communications.

8.6 Comparison with NASA's LLCD Measurement Data

During the LLCD mission a downlink data rate of 622 Mbit/s with a 16-PPM was achieved [8]. The simulation results [4, 5] validate the measurement for all background noise levels better than $K_b = 8.00$.

We would like to mention that until 2022 only the lunar laser communication demonstration (LLCD) by NASA and ESA [8, 9] for Earth-Moon missions have been untertaken. Consequently, the DSOC simulations and the investigation for Earth-Mars missions are even more important for the future. Up to now no real-time experiment for a Earth-Mars mission is available, although, DSOC research has been conducted during Earth-Mars missions. However, there is real time experiment in which laser communication was used for Earth-Moon missions by NASA and ESA as specified on page 22 of the document issued by NASA titled "The Lunar Laser Communication Demonstration (LLCD)" [8]. Events gathered from NASA and ESA projects towards the achievement of a real time experiment for DSOC Earth-Mars mission include:

The planned NASA project MTO (Mars Telecommunications Orbiter) with Laser Communications Demonstration beside X-band and Ka-band RF transmission (this has been canceled for now);

Another project is the NASA Psyche Project in cooperation with ESA planned for 2022–2027, it is ongoing, and has already delivered the first measurement results.

Consequently, we have used a similar downlink DSOC simulation method to compare the results of NASA and ESA in their Earth-Moon downlink lunar laser communication with the closest distance range of 363,253 km and the farthest distance of 405,861 km. Simulation was done for LLCD @ 622Mbps using an average transmit power of 0.5 W, a 16-PPM and a code rate of 0.5 as specified on page 5 of document issued by NASA titled "The Lunar Laser Communication Demonstration (LLCD)" [8].

Below Figures 8.18, 8.19, 8.20, and 8.21 show the slot signal noise ratio, bit signal noise ratio, symbol error probability, and symbol error probability against distance during the downlink simulation of the lunar communication link using a similar simulation method of the DSOC, with assumed average background noise photon rate K_b als varying parameters. Although the Earth-Moon link is limited in distance range, compared to a typical DSOC link, the downlink throughput from the simulation is 622 Mbps to 1230 Mbps depending on the background noise factor K_b as shown in Fig. 8.22 below also in comparison with Table 8.8.

Table 8.8 is an extract from a presentation of NASA LLCD which shows the project accomplishment with a real time experiment conducted for Earth-Moon mission by NASA

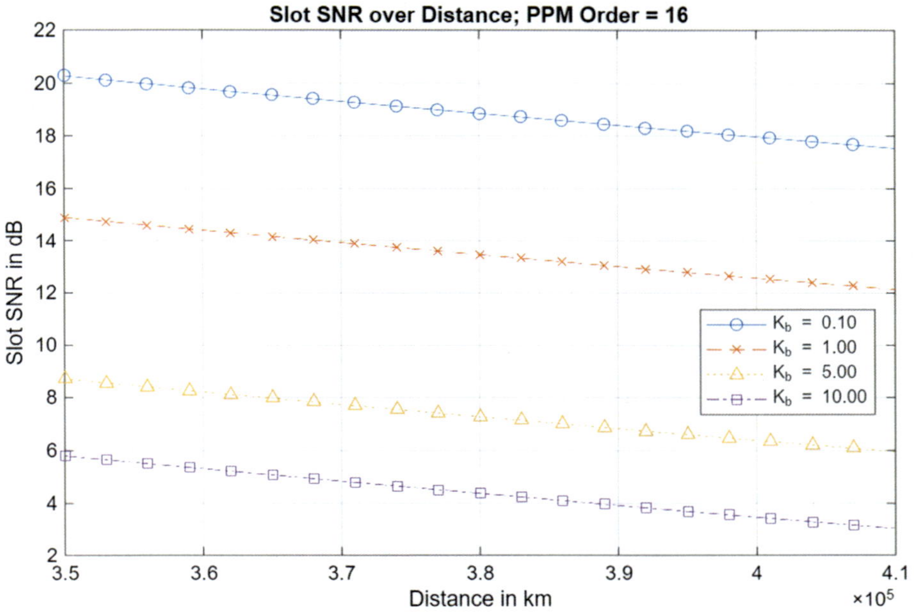

Fig. 8.18 LLCD Slot SNR signal noise ratio versus distance

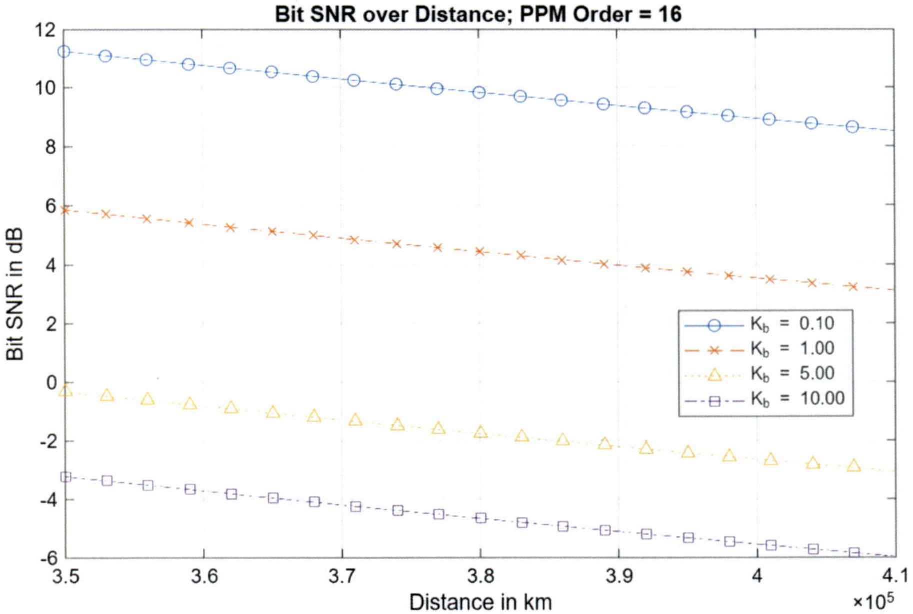

Fig. 8.19 LLCD Bit SNR signal noise ratio versus distance

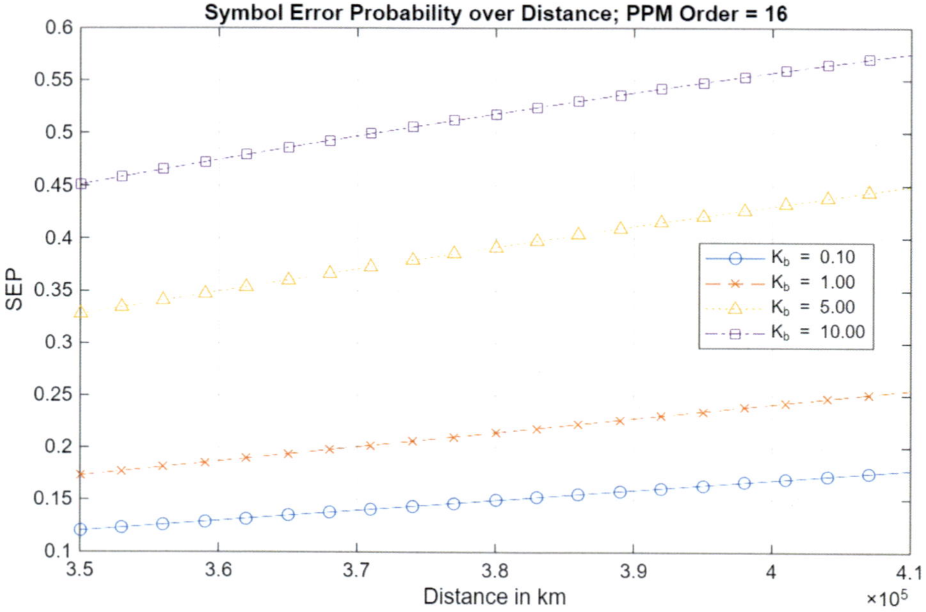

Fig. 8.20 LLCD SEP symbole error probability versus distance

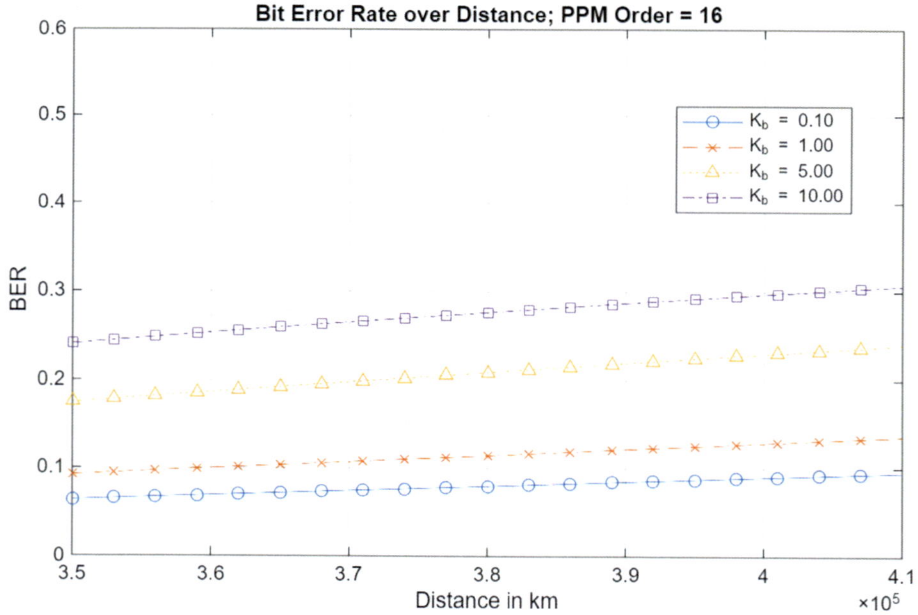

Fig. 8.21 LLCD BER bit error rate versus distance

Table 8.8 Downlink throughput of NASA real time experiment for LLCD

LLCD test	File size (MB)	Duration (s)	Throughput (Mbps)
How high the moon (mp3)	2	0.013	1,230.77
Dark side of the Moon (CD)	451.5	5.8	622.76
Apollo 13	36.800	474	621.10
U.S. Library of congress	208,000,000	2,674,800	622.10

and ESA. The achieved downlink throughput in the real time experiment was from 621.10 Mbps to 1230.77 Mbps depending on sky radiance variation and elevation. This is comparable to the average value of 1230.00 Mbps (Table 8.8) determined from the downlink simulation of Earth-Moon link in Fig. 8.22 depending on sky radiance variation. This also validates the results from the downlink simulation done for the DSOC. The downlink data throughput is dependent on PPM-order, code rate, and existing sky radiance (background noise) at the receiver terminal.

Table 8.8 shows downlink throughput achievements from "The Lunar Laser Communication Demonstration (LLCD) @ 622Mbps" issued by NASA [8].

8.7 Comparison with NASA/ESA's Psyche Mission Measurement Data

During the NASA's Psyche mission in cooperation with ESA, a data rate of 267 Mbit/s was achieved on December 11, 2023, from 0.21 AU or 31 million kilometers away, marked by the red bullet in the Fig. 8.23. On April 8, 2024, the spacecraft laser transceiver transmitted 25 Mbit/s over 1.51 AU or 226 million kilometers to the Earth ground station [6], marked by the blue square in Fig. 8.23.

Due to fact that detailed measurement data for background noise is unknown, in the simulation the background noise factor $K_b = 0.01$ is assumed, even though the background noise levels could be different at different points in time and different space positions during the Psyche mission.

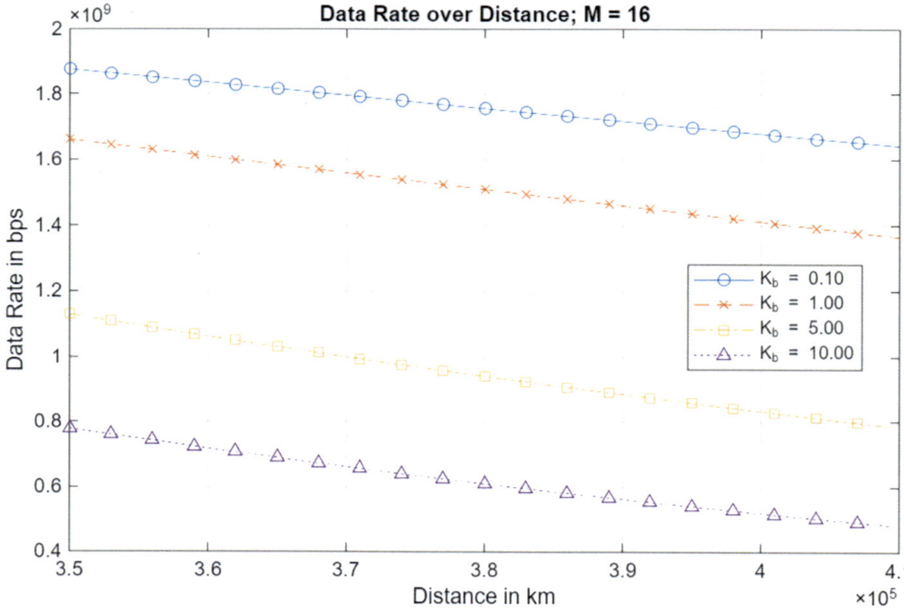

Fig. 8.22 Comparison with the measured data rate 622 Mbit/s for 16 PPM order with NASA Lunar Laser Communication Demonstration

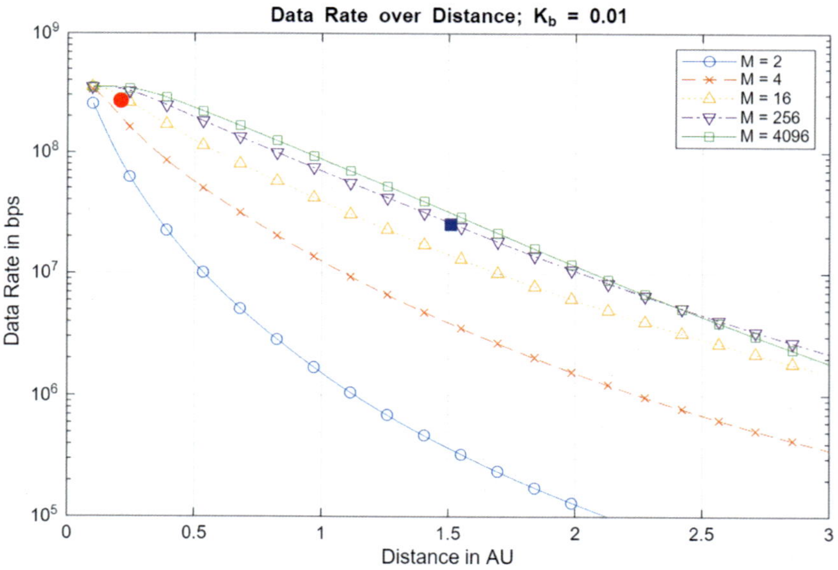

Fig. 8.23 Comparison with the measured data rates for different PPM orders of the NASA/ESA Psyche Mission ●: 267 Mbps at 0.21 AU on December 11, 2023; ■: 25 Mbps at 1.51 AU on April 11, 2024

Never the less, these measurement data rates approximately validate the simulation results [2–4] for different background noise levels and pulse position modulation (PPM) orders.

8.8 NASA/ESA's Artemis II Optical Communications System O2O

From April 1, 2026 - April 11, 2026, during NASA's Artemis II mission, partially in cooperation with ESA, a laser-based system called O2O (Orion Artemis II Optical Communications System) onboard ORION spacecraft was used to transmit data rates up to 260 Mbit/s [11] by using a PPM order of 32 and a 4-inch telescope for the downlink from lunar orbits to the ground station. In comparison, the Psyche mission uses PPM orders 16, 32, 64, and 128 for the DSOC experiments. Therefore the published NASA data [11] somehow confirms the Psyche Mission DSOC measurements and simulation results in Fig. 8.23.

8.9 Conclusions

The simulation methods for free space optical communications (FSOC) and deep space optical communications (DSOC) presented in the last chapters can be utilized to be compared with the measurement data between the near-Earth GEO satellite ARTEMIS, between the Lagrange point L5, and the optical ground station OGS. Especially, the valuable measurement data of ESA's ARTEMIS SILEX project are used to verify, not only the link budget calculation, but also the critical scintillation estimation with feasible parameters.

The measurement data of the NASA LLCD mission, the NASA/ESA Psyche mission, and the data rate achieved by O2O onboard the ORION of ARTEMIS II Mission are compared with the simulation methods explained in this book and verify the simulation models. This encourages the author to further develop the model and continue to investigate the challenging FSOC and DSOC problems.

References

1. Tobias Siegel, Shun-Ping Chen: Investigations of Free Space Optical Communications. Wireless Personal Communications (2021) 116, 475–490. https://doi.org/10.1007/s11277-020-07724-1
2. Shun-Ping Chen: Investigations of free space and deep space optical communication scenarios. CEAS Space Journal (2021). https://doi.org/10.1007/s12567-021-00388-y
3. Shun-Ping Chen: Performance analysis of near earth, lunar and interplanetary optical communication links. Optical and Quantum Electronics (2022) 54. https://doi.org/10.1007/s11082-022-03987-z
4. Emmanuel Domfeh Aboagye, Shun-Ping Chen: Deep space optical communications (DSOC) downlink simulation with varying PPM order. Optical and Quantum Electronics (2021) 53. https://doi.org/10.1007/s11082-021-03232-z

5. F. Schüßler, T. Stieglitz, P. Thielke, S.-P. Chen (Supervisor): FSOC (Free Space Optical Communications) und DSOC (Deep Space Optical Communications). IWP Team Research Project (in German). Faculty of Electrical Engineering and Information Technology, h-da Darmstadt University of Applied Sciences. June 19, 2024.
6. NASA, "NASA's Optical Comms Demo Transmits Data Over 140 Million Miles". (2024), URL: https://www.nasa.gov/missions/psyche-mission/nasas optical-comms-demo-transmits-data-over-140-million-miles/ (Retrieved on May 15, 2024).
7. Zoran Sodnik, Clemens Heese, Pantelis-Daniel Arapoglou, Klaus-Jürgen Schulz, Igor Zayer, Robert Daddato, Stefan Kraft, "Deep-Space Optical Communication System (DOCS) for ESA's Space Weather Mission to Lagrange Orbit L5". IEEE International Conference on Space Optical Systems and Applications (ICSOS), November 14–16, 2017.
8. D. Cornwell, "The Lunar Laser Communication Demonstration (LLCD)". LLCD Mission Manager, NASA GSFC, Presentation to the JHU Aerospace Affinity Group. June 11, 2014.
9. Z. Sodnik, H. Smit, M. Sans, D. Giggenbach, P. Becker, R. Mata-Calvo, C. Fuchs, I. Zayer, M. Lanucara, K.-J. Schulz, J. Widmer, F. Arnold, M. Mosberger, A. Alonso, I. Montilla, "Results from a Lunar Laser Communication Experiment between NASA's LADEE Satellite and ESA's Optical Ground Station", Proc. International Conference on Space Optical Systems and Applications (ICSOS) 2014, S2-1, Kobe, Japan, May 7-9 (2014).
10. Z. Sodnik and R. Czichy, "Design Data Summary of the ESA Optical Ground Station for In-Orbit Check-Out of Laser Communication Payloads and for the Observation and Registration of Space Debris". ESA Specifications for the Artemis OGS OPALE project. Ref.: TEC-MMO/267/ZS. May 2, 2005.
11. NASA: http://www.nasa.org/technology/space-comms/o2o/

9 Conclusions

After the introduction and the motivations to use free space optical communications (FSOC) and deep space optical communications (DSOC), the system model for the simulation of free space optical communications is briefly summarized in the Chap. 2. The simulation model includes all indispensable components like transmitter, receiver, transmission medium which is in fact extremely complex by considering the different, dynamically changing atmosphere layers and leads to disturbances like attenuation, scintillation, beam wandering etc.

Due to the free space wave propagation, beam divergence depends strongly on the gain of the transmitter and receiver telescopes. Together with the maximally achievable transmitter power, the lowest detectable received power and total attenuation loss, the signal-to-noise ratio SNR or signal-to-interference-and-noise ratio SINR can be calculated, and the link budget in free space optical communications can be determined.

Different modulation schemes like OOK, BPSK, QPSK, 16-QAM, or arbitrary M-QAM have been introduced in Chap. 2, depending strongly on the transmission channel conditions or the SINR.

Very important factors for free space optical communications and deep space optical communications are the near-Earth atmospheric layers with their dynamic changing weather conditions, turbulences, and scintillations affecting SINR. These effects have been discussed more in details in Chap. 3.

Based on the Shannon Channel Capacity Limit formula, the maximally achievable transmission bitrates or throughput can be estimated. Beside the payload data, also the redundant frame overhead for synchronization and channel coding or FEC forward error correction are used. But in this book we focus only on the total throughput.

Especially in the very critical transmission conditions for deep space optical communications, the signal-to-interference-and-noise ratio SINR can be further improved by applying

S.-P. Chen, *Fundamentals of Free Space and Deep Space Optical Communications*,
https://doi.org/10.1007/978-3-032-35826-4_9

amplifiers and sensitive photon counters. Different transmitters and receivers have been also discussed in Chap. 4.

In Chap. 5 the different scenarios for the terrestrial, i.e. short distance free space optical communications, have been demonstrated by using 16-QAM and DPSK with smaller transceiver telescopes.

Fortunately many demonstration projects have been conducted by NASA and ESA on near-Earth free space optical communications, especially GEO-stationary and lunar laser optical communications, in the future also interplanetary mission. Our simulation results basically use the published project parameters of theses demonstration projects. The simulation results match the measurement data of the ESA ARTEMIS Satellite measurement data (downlink OPALE to OGS and uplink OGS to OPALE) pretty well. These are presented in Chap. 6.

For deep space optical communications the channel conditions are extremely critical with extremely low SINR leading to a low transmission throughput. For such conditions higher order modulation schemes like M-QAM, QPSK cannot be utilized. Only PPM pulse position modulation schemes can survive and can achieve acceptable bitrates in combination with appropriate channel coding methods (Chap. 7).

In Chap. 8 the simulation results are presented for lunar and deep space optical communications. Comparisons with measurement results by NASA and ESA, like LLCD Lunar Laser Communications Demonstration project, the Psyche Mission with DSOC and the recent ARTEMIS II mission with O2O (Orion Artemis II Optical Communications System), shows a good agreement of simulated and measured data.

Hopefully this book will contribute to the FSOC and DSOC research, and help engineers, and students as future engineers, to understand opportunities and challenges of Free Space Optical Communications and Deep Space Optical Communications, and also understand the principles and methodology, and possibly also motivate them to join in the FSOC and DSOC research and development community.

Frequency Bands

A

A.1 IEEE Frequency Bands

The following tables show the often used IEEE frequency bands with special abbreviations, frequency bands and wavelength ranges (Table A.1).

Table A.1 IEEE frequency bands

Abbreviation	Frequency band	Wavelength range
HF	0.003–0.03 GHz	10–100 m
VHF	0.03–0.3 GHz	1–10 m
UHF	0.3–1 GHz	300–1000 mm
L band	1–2 GHz	150–300 mm
S band	2–4 GHz	75–150 mm
C band	4–8 GHz	37.5–75 mm
X band	8–12 GHz	25–37.5 mm
Ku band	12–18 GHz	16.7–25 mm
K band	18–27 GHz	11.1–17 mm
Ka band	27–40 GHz	7.5–11.1 mm
V band	40–75 GHz	4–7.5 mm
W band	75–110 GHz	2.7–4 mm
mm waves	100–300 GHz	1–2.7 mm
THz band	300 GHz to 3 THz	0.1–1 mm
Infrared	3–480 THz	0.625–100 μm
Visible light	480–680 THz	0.44–0.625 μm

S.-P. Chen, *Fundamentals of Free Space and Deep Space Optical Communications*,
https://doi.org/10.1007/978-3-032-35826-4

The IEEE frequency bands and wavelength ranges of the most used electromagnetic waves in information and communication technologies, extended in this table with the terahertz (THz) band, infrared IR-band and visible light.

A.2 ITU Frequency Bands

The following tables shows the ITU frequency bands with special abbreviations, frequency bands and wavelength ranges (Table A.2).

ITU frequency bands and wavelength ranges of the most used electromagnetic waves in information and communication technologies with the special abbreviations, frequency bands, and wavelength ranges.

Table A.2 ITU frequency bands

ITU No.	Abbreviation	Frequency bands	Wavelengths
1	ELF	3–30 Hz	100,000–10,000 km
2	SLF	30–300 Hz	10,000–1000 km
3	ULF	300–3000 Hz	1000–100 km
4	VLF	3–30 kHz	100–10 km
5	LF	30–300 kHz	10–1 km
6	MF	300–3000 kHz	1000–100 m
7	HF	3–30 MHz	100–10 m
8	VHF	30–300 MHz	10–1 m
9	UHF	300–3000 MHz	1–0.1 m
10	SHF	3–30 GHz	100–10 mm
11	EHF	30–300 GHz	10–1 mm
12	THz, THF	300–3000 GHz	1–0.1 mm
	Infrared	3–480 THz	0.625–100 μm
	Visible light	480–680 THz	0.44–0.625 μm

A.3 EU NATO Frequency Bands

The following tables shows the EU NATO frequency bands with special abbreviations, frequency bands and wavelength ranges (Table A.3).

Table A.3 EU NATO bands

IEEE	Band	EU/NATO	Band
L	1–2 GHz	D	1–2 GHz
S	2–4 GHz	E	2–3 GHz
C	4–8 GHz	F	3–4 GHz
X	8–12 GHz	G	4–6 GHz
Ku	12–18 GHz	H	6–8 GHz
K	18–27 GHz	I	8–10 GHz
Ka	27–40 GHz	J	10–20 GHz
V	40–75 GHz	K	20–40 GHz
W	75–110 GHz	L	40–60 GHz
mm	110–300 GHz	M	60–100 GHz
Terahertz	0.3–3 THz	Terahertz	0.3–3 THz
Infrared	3–480 THz	Infrared	3–480 THz
Visible light	480–680 THz	Visible light	80–680 THz

A.4 Optical Fiber Communication Frequency Bands

Table A.4 summarizes the classical optical fiber communication bands. Many of the optical fiber communication system components are directly or modified for free space optical communications or deep space optical communications, like transmitters, receivers, EDFA Erbium doped fiber amplifiers etc..

Beside these bands, also wavelengths of 850 nm, and 1064 nm are used especially for deep space optical communications with high-power lasers.

Table A.4 ITU optical fiber communication bands (1260–1675 nm)

Band	Name	Wavelength (nm)	Typical applications
O-band	(Original)	1260–1360	Low dispersion
E-band	(Extended)	1360–1460	Higher attenuation
S-band	(Short-wavelength)	1460–1530	Passive optical network
C-band	(Conventional)	1530–1565	Loss minimum, WDM/EFDA
L-band	(Long-wavelength)	1565–1625	Capacity extension
U-band	(Ultra-long-wavelength)	1625–1675	OTDR, monitoring

Glossary

16QAM Quadrature Amplitude Modulation with 16 symbols
32QAM Quadrature Amplitude Modulation with 32 symbols
64QAM Quadrature Amplitude Modulation with 64 symbols
128QAM Quadrature Amplitude Modulation with 128 symbols
256QAM Quadrature Amplitude Modulation with 256 symbols
APD (Geiger-Mode) Avalanche Photodiode
A_{eff} Effective receive area
ARTEMIS SILEX ESA Satellite (Advanced Relay and Technology Mission) with Laser Communication Terminal SILEX (2001-2014)
ARTEMIS to OGS ESA Satellite (Advanced Relay and Technology Mission) to Optical Ground Station (2001-2014)
AU Astronomical Unit, distance between Earth and Sun, about 150 million km
BEP Bit error probability
BER Bit error rate
BPSK Binary Phase Shift Keying
BR Bitrate
B_w Bandwidth
C Band 4 GHz - 8 GHz band
CR Code Rate
CWDM Coarse Wavelength Division Multiplex
DBR Distributed Bragg Reflexion Laser
DPPS Diode-Pumped Solid-State Laser
$\Delta\lambda_{nr}$ Optical filter bandwidth in μm
DFB Distributed Feed Back Laser
DGT DOCS Ground Terminal
DL Downlink

S.-P. Chen, *Fundamentals of Free Space and Deep Space Optical Communications*,
https://doi.org/10.1007/978-3-032-35826-4

DLR Deutsches Zentrum für Luft- und Raumfahrt (German Aerospace Center)
DOCS L5 ESA Project Deep Space Optical Communications System Lagrange 5
DP-QPSK Dual-Polarization QPSK
D_{rx} Receiver aperture diameter
D_{tx} Transmitter aperture diameter
DSOC Deep Space Optical Communications
DSP Digital Signal Processing
DSN Deep Space Networks
DST DOCS Space Terminal
DUT Device under test
DWDM Dense Wavelength Division Multiplex
ESTRACK ESA's global tracking network at ESOC comprising 6 stations in 6 countries
Eb/No Energy per bit to noise power spectral density ratio
EDFA Erbium doped fiber amplifier
Er/Yb Erbium/ytterbium doped optical fiber or high power fiber laser
ESA/ESOC European Space Agency, European Space Operations Centre
η Optical efficiency or quantum efficiency
η_{Atm} Atmospheric transmittance
η_{FS} Free space propagation efficiency
η_P Pointing efficiency
η_R Receiver telescope efficiency
η_T Transmitter telescope efficiency
EU European Union
FEC Forward Error Correction
r_0 Fried parameter
FOV Field of View
FSL Free Space Loss
FSOC Free Space Optical Communications
γ_{FOV} Receiver field of view measured in steradians (sr)
GEO Geostationary Earth Orbit
g_R Receiver telescope gain
g_T Transmitter telescope gain
h Planck's constant
H_b Sky radiance or background irradiation energy density
HDTV High Definition Television
HPD Hybrid Photomultiplier Detector
IAF International Astronautical Federation
I_b Bulk leakage current
IEEE Institute of Electrical and Electronics Engineers
InGaAsP Indium gallium arsenide phosphide laser, a III/V semiconductor laser

IOL Inter-orbit link
IPsec Network layer or layer 3 protocol for security applications
I_b Bulk leakage current
I_s Surface leakage current
ISL Inter-satellite link
ISO International Standardization Organization
ITU International Telecommunication Union
ITU-T ITU Recommendations for Telecommunication Systems and Applications
ITU-R ITU Radiocommunication Standards
JAXA Japan Aerospace Exploration Agency
k Bits per second
K Band 18-21 GHz band
K_b Background photon rate
k_B Boltzmann constant
k_{eff} Ratio of the hoe and electron ionization coefficients
K_s Signal photon rate
Ka Band 27-40 GHz band
Ku Band 12-18 GHz band
λ Wavelength
Laser Light Amplification by Stimulated Emission of Radiation
L Band 1-2 GHz band
LCT Laser Communication Terminal
LDPC/Turbo Low Density Parity Check coding, Turbo coding
LF Band Low Frequency Band
LLCD Lunar Laser Communications Demonstration
L_{margin} Link margin
LOS/LoS Line of Sight propagation condition or environment
L_{point} Pointing loss
L_R Rayleigh distance
L_{scint} Scintillation loss
L_{trans} Transmission loss
LEO Low Earth Orbit
MEO Medium Earth Orbit
M or m PPM order or QAM order
M-PPM or m-PPM M-nary Pulse Position Modulation
M-QAM or m-QAM M-nary Quadrature Amplitude Modulation
MOPA Master Oscillator Power Amplifier
M_r Relative air mass
MTO NASA's Mars Telecommunications Orbiter
NASA National Aeronautics and Space Administration
Nd:YAG Neodymium-doped yttrium aluminium garnet laser)

NATO North Atlantic Treaty Organization
NICT National Institute of Information and Communications Technology, Japan
NIR Near Infrared
NLOS Non Line of Sight propagation condition
OGS Optical Ground Station
OPALE Optical Payload Laser Experiment onboard ARTEMIS
OOK On-Off-Keying
O2O Orion Artemis II Optical Communications System
OWC Optical Wireless Communications
PAPR Peak to Average Power Ratio
P_{avg}, P_T Transmitter laser power
P_b or P_{back} Received background power
P_R Received laser power
PIN-Diode Semiconductor photodiode consisting of p-i-n junctions
PLL Phase Locked Loop
PMT Photomultiplier Tubes
PPM Pulse Position Modulation
PRR Pulse repetition rate
QAM Quadrature Amplitude Modulation
QCL Quantum Cascade Lasers
QDL Quantum Dot Lasers
4PSK/QPSK Quadrature Phase Shift Keying modulation technique
c_n^2 Refractive index structure parameter, directly related with Rytov variance σ_I^2
RF Radio Frequency
R_L Load resistance
RX Receiver
S Band 2-4 GHz band
SEP Sun-Earth-Probe angle
SEP Symbol error probability
SER Symbol error rate
Si-APD Silicon Avalanche Photodiodes
SILEX Semiconductor-Laser Intersatellite Link Experiment onboard ARTEMIS
SiPM Silicon Photomultipliers
SINR Signal to Interference and Noise Ratio
SNR Signal to Noise Ratio
SNSPD Superconducting Nanowire Single Photon Detector
SSA ESA's Space Situational Awareness Program
σ_{dc} Dark current noise
σ_{bg} Background noise
σ_{sn} Shot noise

σ_{th} Thermal noise
σ_I^2 Rytov variance, directly related with the refractive index structure parameter c_n^2
SWE ESA's Space WEather L5 Mission
T_{atm} Atmospheric transmittance
TEM_{00} Transverse Electro-Magnetic mode, so-called fundamental mode
TESAT Tesat-Spacecom GmbH & Co. KG, Backnang, Germany
θ_{FOV} Planar angular detector field of view in radians
θ Atmospheric transmittance
THF Band Tremendously High Frequency Band
TLS/SSL Transport layer protocol for security applications or Secure Socket Layer
T_r Electronic system noise temperature
T_s Signal slot width
T_{zenith} Atmospheric zenith transmittance
TX Transmitter
UHF Band Ultra High Frequency Band
UL Uplink
UV Ultraviolet
VHF Band Very High Frequency Band
VLF Band Very Low Frequency Band
w_0 Beam radius or half spot size at the transmitter
$w(L)$ Beam radius or half spot size at the distance L to transmitter
WDM Wavelength Division Multiplex

Index

S.-P. Chen, *Fundamentals of Free Space and Deep Space Optical Communications*, https://doi.org/10.1007/978-3-032-35826-4

FSC
www.fsc.org
MIX
Papier | Fördert gute Waldnutzung
FSC® C083411

Kolibri 360 GmbH
Ferdinand-Jühlke-Straße 7
99095 Erfurt, Deutschland
produktsicherheit@kolibri360.de